W0091867

OXY-FUEL COMBUSTION

THE NOX AND COAL IGNITION REACTIONS

ENERGY SCIENCE, ENGINEERING AND TECHNOLOGY

Additional books in this series can be found on Nova's website under the Series tab.

Additional E-books in this series can be found on Nova's website under the E-book tab.

ENERGY SCIENCE, ENGINEERING AND TECHNOLOGY

OXY-FUEL COMBUSTION

THE NOX AND COAL IGNITION REACTIONS

MASAYUKI TANIGUCHI

AND

KENJI YAMAMOTO

Nova Science Publishers, Inc.

New York

For permission to use material from this book please contact us:
Telephone 631-231-7269; Fax 631-231-8175
Web Site: http://www.novapublishers.com

NOTICE TO THE READER

The Publisher has taken reasonable care in the preparation of this book, but makes no expressed or implied warranty of any kind and assumes no responsibility for any errors or omissions. No liability is assumed for incidental or consequential damages in connection with or arising out of information contained in this book. The Publisher shall not be liable for any special, consequential, or exemplary damages resulting, in whole or in part, from the readers' use of, or reliance upon, this material. Any parts of this book based on government reports are so indicated and copyright is claimed for those parts to the extent applicable to compilations of such works.

Independent verification should be sought for any data, advice or recommendations contained in this book. In addition, no responsibility is assumed by the publisher for any injury and/or damage to persons or property arising from any methods, products, instructions, ideas or otherwise contained in this publication.

This publication is designed to provide accurate and authoritative information with regard to the subject matter covered herein. It is sold with the clear understanding that the Publisher is not engaged in rendering legal or any other professional services. If legal or any other expert assistance is required, the services of a competent person should be sought. FROM A DECLARATION OF PARTICIPANTS JOINTLY ADOPTED BY A COMMITTEE OF THE AMERICAN BAR ASSOCIATION AND A COMMITTEE OF PUBLISHERS.

Additional color graphics may be available in the e-book version of this book.

LIBRARY OF CONGRESS CATALOGING-IN-PUBLICATION DATA

Taniguchi, Masayuki, 1958-
 Oxy-fuel combustion : the NOx and coal ignition reactions / Masayuki
Taniguchi and Kenji Yamamoto.
    p. cm.
 Includes index.
 ISBN 978-1-61668-979-7 (softcover)
 1. Coal gasification. 2. Flue gases--Combustion. 3. Gases at high
temperatures. 4. Nitrogen oxides. 5. Carbon sequestration. I. Yamamoto,
Kenji, 1969- II. Title.
 TP759.T36 2009
 621.31'21320286--dc22
                        2010024037

Published by Nova Science Publishers, Inc. † New York

CONTENTS

PREFACE

It has been predicted that by 2030 the world's population will be about 1.6 times that of 2004. The energy consumption will increase by 60% by 2030. This new book discusses the Wind turbine generation and photovoltaic power generation which are expected to increase ten-fold by 2030, but their combined percentage of the generation will only total 7%. The expectation for nuclear power generation is large, but under present conditions, the amount of natural uranium resources will have a share for only 80 years. Development of the fast breeder reactor is being pushed forward, but it is not expected to be commercial before about 2050. Therefore, it will continue to be necessary to depend on fossil fuel. Carbon dioxide capture and storage (CCS) techniques for thermal power plants will also surely become important.

NOMENCLATURE

a	constant decided by the shape of the burner and furnace
b	constant
C	amount of fuel supply (W/burner)
CHi	hydrocarbon radicals
C_{min}	minimum fuel supply for stable combustion (W/burner)
L	distance between the burner nozzle and sampling port (mm)
Ox(OH)	oxidative radicals, such as OH
Q	heat of combustion which flows in the recirculation region (W/burner)
Q_{in}	heat flows in primary air from the recirculation region (W/burner)
Q_{min}	minimum heat that should flow in the primary air from the recirculation region, to form a stable flame (W/burner)
Q_{rad}	radiant heat loss to the furnace wall from the recirculation region (W/burner).
QradC	radiant heat loss to the caster wall from the recirculation region (W/burner).
QradL	heating rate from surroundings (W/burner)
QradW	radiant heat loss to the water wall from the recirculation region (W/burner).
Qvm	volatile content of coal as the calorific value
SRgas	gas phase stoichiometric ratio

SRin	inlet stoichiometric ratio
XN (NHi、HCN)	nitrogenous radicals
α	fraction of heat of combustion that flows in the recirculation region and that flows in the primary air (carrier gas); 1- α means the fraction of heat loss from the flame to the furnace wall
β	fraction of the area covered with the caster judged from the viewpoint of the burner nozzle
[CO]	mole fraction of carbon monoxide
[CH_4]	mole fraction of methane
[CHi]	mole fraction of hydrocarbon radicals
[H_2]	mole fraction of hydrogen
[O_2]	mole fraction of oxygen
[O_2]$_0$	mole fraction of oxygen in combustion supporting gas
[OH]	mole fraction of OH
[OH $_{eq}$]	equilibrium mole fraction of OH
[TR]	mole fraction of tracer
[TR]$_0$	mole fraction of tracer in combustion supporting gas

Chapter 1

INTRODUCTION

At COP13 held in Bali in December-2007, the framework for cubing-global warming were taken after 2013. The development of the technologies that are necessary for cubing-global warming must be accelerated through worldwide cooperation.

The IEA (International Energy Agency) has predicted that by 2030 the world's population will be about 1.6 times that of 2004. The energy consumption will increase by 60% by 2030. Figure 1 shows predicted energy consumption in 2030 for various uses (cited in [1]). The percentage consumed in power generation is expected to be the largest and will be 45%.

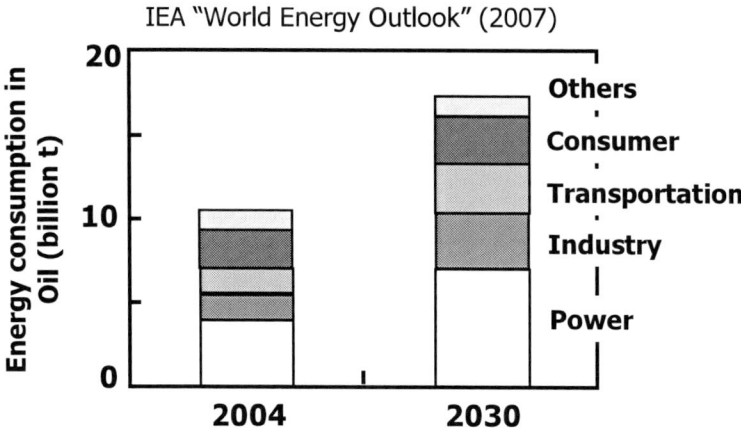

Figure 1. The prediction of worldwide energy demand (from IEA "World Energy Outlook: 2007" and cited in [ref.1]).

Wind turbine generation and photovoltaic power generation are expected to increase ten-fold by 2030, but their combined percentage of the generation will only total 7% [1]. The expectation for nuclear power generation is large, but under present conditions, the amount of natural uranium resources will have a share for only 80 years [1]. Development of the fast breeder reactor is being pushed forward, but it is not expected to be commercial before about 2050 [1]. Therefore, it will continue to be necessary to depend on fossil fuel. Carbon dioxide capture and storage (CCS) techniques for thermal power plants will surely become important.

Oxy-fuel combustion (CO_2/O_2 combustion) is one of the promising technologies for CCS in coal power plants. Nitrogen is separated beforehand from air to be used for combustion and coal is burnt using the oxygen-containing fraction. The major components of the exhaust gas are H_2O and CO_2. CO_2 can be collected without having to separate it from the nitrogen and H_2O of the exhaust gas. A part of the exhaust gas is circulated in a system to control combustion conditions such as flame temperature. An example of the oxy-fuel combustion system [2] is shown in figure 2. Oxy-fuel combustion technology is applied for pulverized coal combustion. Nitrogen is separated by the ASU (air separation unit) from air to be used for combustion. Only oxygen is supplied to a boiler. Steam is formed by heat exchange with the flame in the boiler. The generated steam drives a steam turbine and generates electricity. The harmful species to atmosphere environment in the combustion exhaust gas are removed by De-NOx, ESP, and De-SOx. The major components of the flue gas are CO_2, H_2O and a little oxygen. Liquid CO_2 is provided after H_2O is removed by compressing the exhaust gas, using a dryer. This system looks like a conventional pulverized coal firing boiler. However, the point that the combustion supporting gas becomes O_2-CO_2-H_2O differs from air combustion systems greatly. CO_2 and H_2O are not always inert towards chemical reactions. It is necessary to determine any influence they may have on the combustion reaction. In addition, specific heats of CO_2 and H_2O are larger than that of nitrogen It is necessary to determine the influence on heat transfer performance. The emissivities of CO_2 and H_2O are also different from that of nitrogen [3].

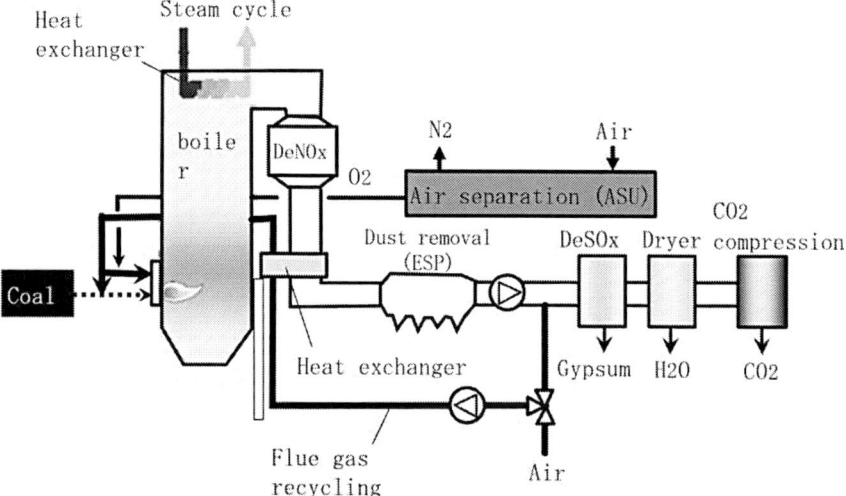

Figure 2. Schematic of a pulverized coal fired oxy-fuel power plant system [ref.2].

In the present investigation, we examined influences on chemical reactions. We focused on the NOx and coal ignition reactions. In the oxy-fuel combustion system, many plants will be developed by using numerical computations, because there are no actual plants yet. As computers become even more sophisticated, their use is increasing for making the numerical analyses needed in designing pulverized coal firing boilers [4-10]. Numerical analyses were first applied to design heat exchanger such as predictions of the steam temperature [4]. Recently, they have been applied to environmental performance factors, such as NOx emission [5-9] and to control of furnace wall corrosion [10].

In this paper, we introduce combustion models for the NOx and coal ignition reactions. We also show examples of case studies for an oxy-fuel combustion system by using the proposed models.

Chapter 2

NOx REACTION MODEL

2.1. A KEY INDEX FOR NOx REDUCTION IN FUEL-RICH CONDITIONS

NOx reduction in flames has been extensively investigated. However, for combustion of coal, NOx performance changes easily with the burning conditions, such as coal properties and coal particle diameter. Boiler design and development cannot be done efficiently if it is necessary to change the method of NOx reduction for each coal property. Then, we proposed a key index to estimate NOx reduction performance.

NOx is reduced both gas phase and solid phase reactions for pulverized coal combustion [5]. We tried to analyze the experimental data by dividing them into the associated gas phase and solid phase reactions. Especially, we focused on the gas phase reaction. In the present study, we proposed a gas phase stoichiometric ratio (SRgas) as the key index.

Figure 3 is a schematic representation of SRgas. Before combustion, all fuel components of pulverized coal are in the solid phase (figure 3a). Pulverized coal particles are surrounded by combustion supporting gas such as air. Inlet stoichiometric ratio (SRin) is often used as an index showing burning conditions. The inlet stoichiometric ratio is generally defined by equation (1) and all the fuel is in solid, liquid and gas phases.

SRin ≡ amount of fuel required for stoichiometric combustion
        /amount of fuel actually supplied   (1)

After coal is ignited, part of the fuel components move from the solid phase to the gas phase by pyrolysis, oxidation, and gasification reactions. The remaining fuel components stay in the solid phase. An image is shown in figure 3b. The gas phase stoichiometric ratio is the index which focuses on the amount of fuel components which moved from the solid phase to the gas phase. The gas phase stoichiometric ratio (SRgas) is defined by equation (2).

SRgas ≡ amount of fuel required for stoichiometric combustion
　　　　/ amount of gasified fuel　　　　　　　　　　　　　　　　　(2)

Here, amount of gasified fuel means both the amount of fuel that has moved from the solid phase to the gas phase by pyrolysis, oxidation, and gasification reactions, and, the amounts of gas and liquid fuel supplied to the combustible mixture. We do not consider the fuel components which are left in the solid phase.

　　　The gas phase reaction rate exceeds the solid phase reaction rate. The inlet stoichiometric ratio is a good index which shows the difference in the burning conditions. We thought a numerical analysis might become easier when solid is removed from the burning mixture and a stoichiometric ratio is defined to consider the gas phase reaction.

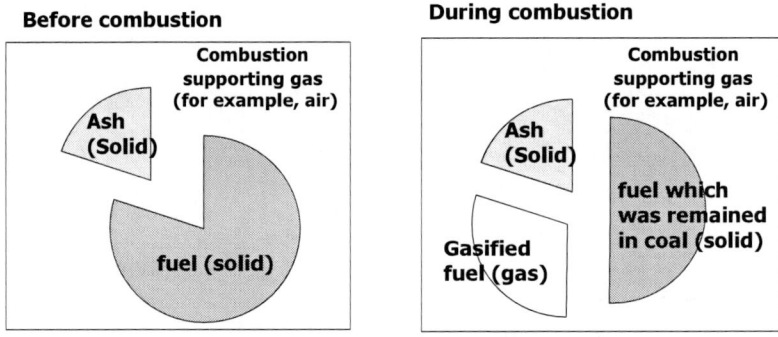

Figure 3. Definition of gas phase stoichiometric ratio (SRgas).

　　　The SRgas can be obtained by analyzing the mass balance of H, C and O in the burning gas. Sometimes, it is difficult to analyze mass balance correctly, because the amount of water is difficult to measure. At this time, SRgas can be obtained approximately as equation (3), too.

$$SRgas = \frac{[O_2]_0/[TR]_0}{[O_2]_0/[TR]_0 - [O_2]/[TR] + 0.5*[CO]/[TR] + 0.5*[H_2]/[TR] + 2*[CH_4]/[TR]} \quad (3)$$

Here, $[O_2]$, $[CO]$, $[H_2]$, and, $[CH_4]$ are O_2, CO, H_2, and, CH_4 concentrations in the burning gas. $[O_2]_0$ is the average concentration of O_2 in the combustion supporting gas. $[TR]$ is the concentration of tracer. N_2, Ar and He are examples of suitable tracers because their amounts will hardly be changed by chemical reactions. In the present study, we used N_2. $[TR]_0$ is the concentration of tracer in the combustion supporting gas.

The relation between the gas phase stoichiometric ratio (SRgas) and NOx concentration was investigated using a drop-tube furnace [11]. A schematic drawing of the drop-tube furnace is shown in figure 4. Pulverized coal and all of the combustion gas were pre-mixed, and then injected to the furnace through a nozzle (inner diameter: 6mm). The nozzle was cooled by water to prevent pyrolysis of coal particles before injection. The nozzle was covered with firebrick and an SiC tube to prevent too much cooling of the injected gas. Flow rate of combustion supporting gas was $0.96Nm^3/h$. Coal feed rate was varied for each experiment from 0.02-0.5 kg/h.

The reaction tube was made of alumina and had an inner diameter and length of 50mm and 1200mm, respectively. Four electric heaters were arranged around the reaction tube. Temperature of each heater could be controlled independently in order to keep the temperature distribution of the tube wall constant. The axial temperature distribution of the heated gas was measured along the center axis of the reaction tube. The difference between the wall and gas temperatures was ±50K. Heating rate of the combustion gas was around 10000 - 15000K/s. The heating rate was estimated by CFD calculations [7,11].

A sampling probe collected all of the burned gas and char. Usually, this probe was set 800mm downstream from the nozzle. The reaction time was around 1.0s. The axial sampling position was varied for some experimental conditions. Cooling water was injected into the probe, and the combustion reactions were quenched.

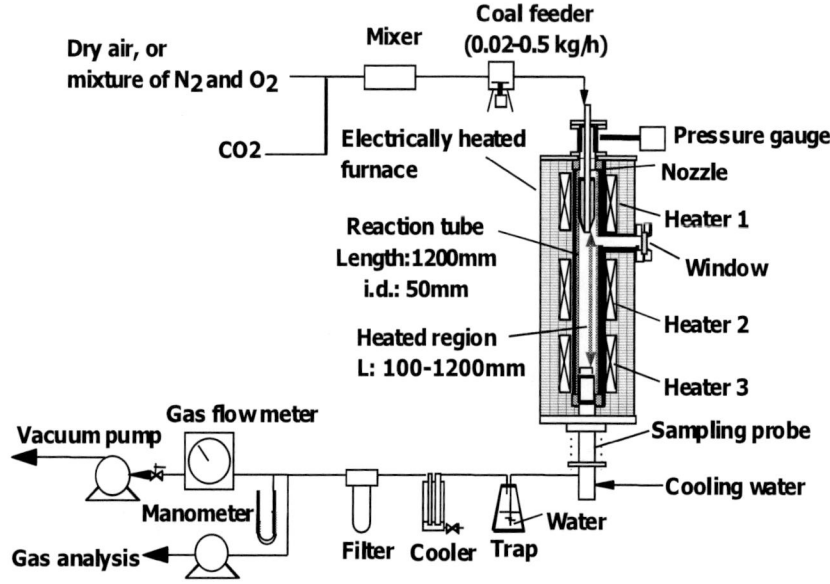

Figure 4. Schematic of the drop-tube furnace [ref.11].

Char was collected by filtration through a 7μm pore paper filter. C, H, N and ash contents in the char were analyzed. Coal burnout was obtained by assuming that the amount of ash remained constant during combustion. Concentrations of HCN and NH_3 were obtained from the concentrations of NH_4^+ and CN^- in the water in a trap and total gas flow rate.

Figure 5 plots NOx concentration characteristics obtained by the drop-tube furnace experiments. The results were obtained for a variety of coal properties, burning temperatures, and compositions of combustion supporting gas. Relationships between both SRin and NOx, and SRgas and NOx are shown in the figure. When the data were analyzed by SRin, NOx concentration varied with experimental conditions. However, when the data were analyzed by SRgas, the difference of NOx concentration became small and NOx was hardly influenced by the burning conditions when SRgas was less than 1.0. We judged that the SRgas was a good index which estimated NOx concentration in fuel-rich conditions. We further thought that the gas-phase NOx reduction was the key reaction when SRgas was smaller than 1.0; that is, the mechanism of the gas reaction did not depend so much on coal properties and there were many common points.

Symbols	Coal	Combustion supporting gas	Temperature	Reaction time
☐	Anthracite	air	1673K	1s
◇	Subbituminous	air	1673K	1s
○	hv bituminous	air	1573-1673K	1s
●	hv bituminous	O2(21vol%)-CO2(79vol%)	1673K	1s
◉	hv bituminous	O2(30vol%)-CO2(70vol%)	1673K	1s

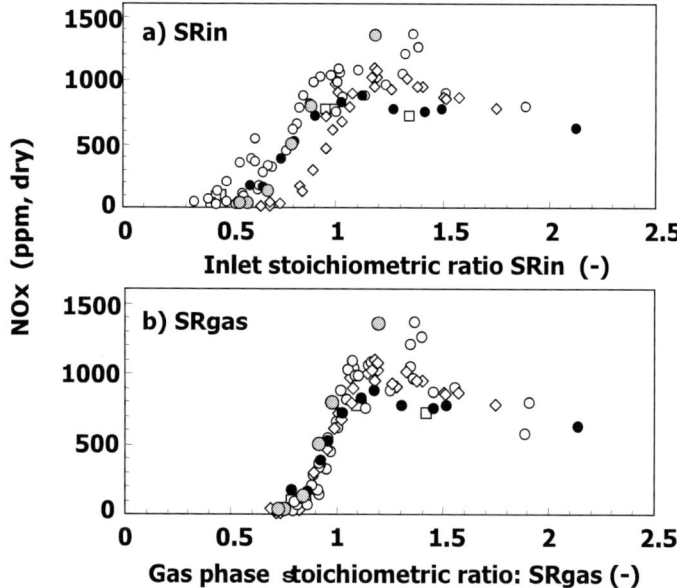

Figure 5. Relationships between NOx concentration for the drop-tube furnace experiments and a) inlet stoichiometric ratio and b) gas phase stoichiometric ratio.

The NOx was also fixed by the gas phase stoichiometric ratio (SRgas) for oxy-fuel combustion. The NOx could be decreased by both air and oxy-fuel combustion, if SRgas could be lowered. Rate of Gasification reactions influences how much SRgas can be reduced in the fuel rich flame. If the rates of gasification reactions (equation (4) and (5)) are large, SRgas in the fuel rich flame can be easily decreased.

$$Char + CO_2 \rightarrow 2CO \tag{4}$$

$$Char + H_2O \rightarrow H_2 + CO \tag{5}$$

Char in coal is oxidised by gasification reactions in the fuel rich flame, because there is little oxygen in a flue gas of fuel Rich flame. Gasification reaction of the (4) is accelerated for oxy-fuel combustion, because CO_2 concentration is very high.

Figure 6 shows the relationship between SRin and coal burnout and the relationship between SRin and SRgas. The experimental results were compared when the same coal burnt with CO_2/O_2 atmosphere and with in the air. Figure 6a shows the relation between SRin and coal burnout. At first, we compared the results when coal was burned at the same oxygen concentration (21vol%) and burning temperature (1673K). Under these experimental conditions, the contribution of oxidation reaction was the same both for air combustion and oxy-fuel combustion. When SRin was small, coal burnout of oxy-fuel combustion was larger than that of air combustion. This was because the contribution of the gasification reaction increased for oxy-fuel combustion, so the amount of gasified fuel increased. When the amount of gasified fuel increased, SRgas decreased easily. If SRgas decreased, NOx concentration decreased easily.

Effect of inlet oxygen concentration was also examined for oxy-fuel combustion (O_2=21, 30vol%). The oxidation reaction was also accelerated by increasing the inlet oxygen concentration. When the inlet oxygen concentration was larger, coal burnouts in the fuel-lean condition (SRgas>1.0) also became larger. When coal burnout increased in the fuel-rich condition (SRgas<1.0), NOx decreased easily. When coal burnout increased in the fuel-lean flame, combustion efficiency was increased easily. Usually, in pulverized coal firing power plants for oxy-fuel combustion, the operating oxygen concentration is larger than that of air combustion; however the burning temperatures for both are almost the same [12]. Therefore, improvement of the coal burnout and reduction of NOx is easy for oxy-fuel combustion.

Effect of burning temperature on coal burnout for oxy-fuel combustion is shown in figure 6. When the burning temperature was increased, the char reaction was accelerated for both fuel-lean and fuel-rich conditions. Only a little nitrogen is contained in the burning gas for oxy-fuel combustion. The reaction rate of thermal NOx is small. Therefore, oxy-fuel combustion is advantageous to reduction of the NOx and improvement of the combustion efficiency when combustion temperature is increased as much as possible.

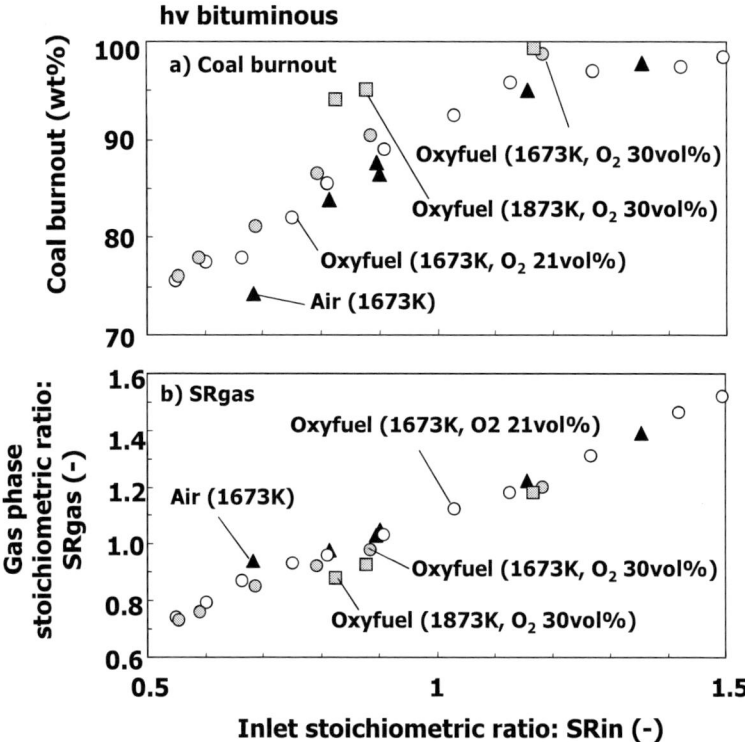

Figure 6. Effects on inlet stoichiometric ratio (SRgas) of a) coal burnout and b) gas phase stoichiometric ratio (SRgas) at the exit of the drop-tube furnace.. The data were obtained from both air combustion and oxy-fuel combustion.

2.2. NOx Reaction Model

An outline of the proposed reaction model is shown in figure 7 [11]. NOx reduction by hydrocarbon radicals (CHi) and char, formation of NOx from the char-N, and the extended Zeldovich mechanism were included. Part (η) of char-N was assumed to become NOx. The value of η was obtained from Tullin et al. [13]. The main feature of the model is the description of the gas phase reaction of NOx reduction in the char burning area. The formation route of CHi is shown in figure 7b. We assumed that CHi was not only formed during pyrolysis of volatile matter, but also during oxidation and gasification of char. In this way, NOx concentration in fuel-rich conditions could be predicted well.

a) Reaction schema

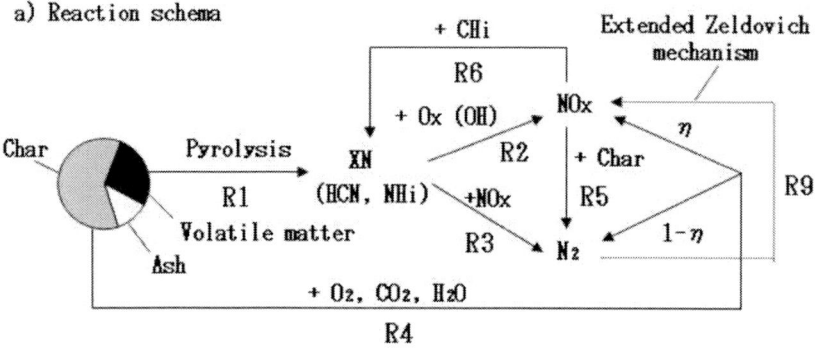

b) Formation route of CHi

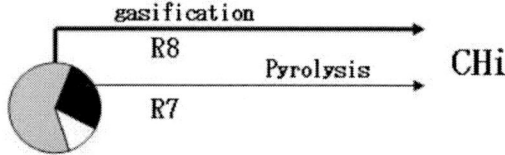

Figure 7. A reaction model for NOx formation and reduction [ref.11].

Char is usually assumed as carbon (C) in the numerical analysis of coal combustion [5]. However, actually, char is not carbon but is a hydrocarbon substance [11]. For example, the nominal composition of char is approximated as $C_8H_{1.04}N_{0.15}$ for a bituminous coal [11] (volatile matter content around 30vol%, dry, ash-free basis). We assumed that part of the hydrogen (H) in the char formed hydrocarbon radials (CHi).

Figure 8 shows the relationship between coal burnout and atomic ratios of H/C and N/C in burning particles in a bituminous coal [11]. We tested the atomic ratios by using the drop-tube furnace [11] and a laboratory scale furnace [14]. H content decreased in the volatile combustion region, but H atoms still remained in the final burnout stage of the char combustion region. In the char combustion region, the H/C ratio was almost constant with coal burnout. The results meant that H atoms were released into the gas phase at almost the same rate as carbon. Figure 9 shows an example of coal burnout and NOx distributions for bituminous coal in the laboratory scale furnace [11]. The grey area in figure 9b shows the volatile combustion region. We estimated that the actual volatile content was 1.5 times larger than that of the proximate analysis. NOx rapidly increased in the primary stage, then reached its maximum value which was as much as 700ppm. And then, NOx rapidly decreased downstream in the char combustion region.

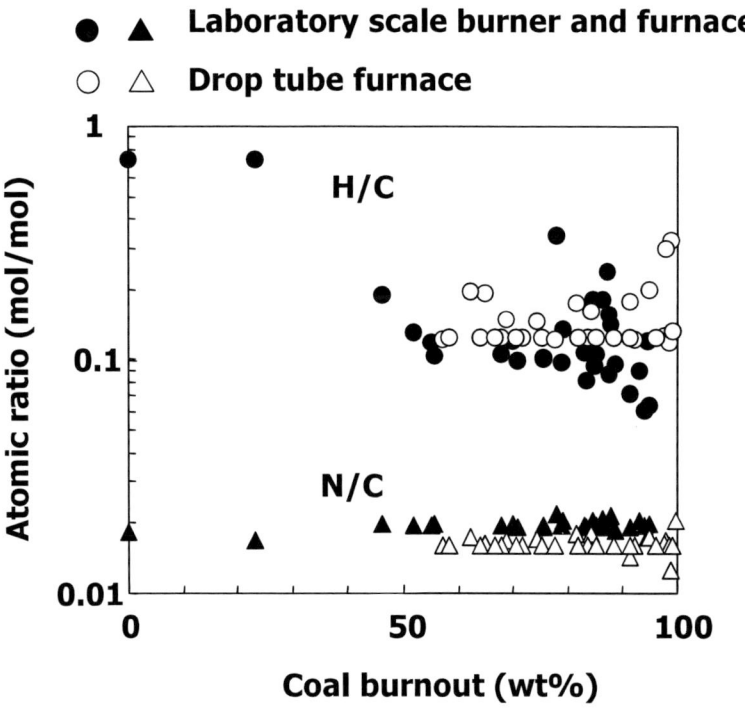

Figure 8. Relationships between coal burnout and atomic ratios (H/C and N/C) in solid particles collected from flames produced by hv bituminous coal. ○, H/C obtained with drop-tube furnace; ●, H/C with laboratory scale furnace; △, N/C with drop-tube furnace; ▲, N/C with laboratory scale furnace [ref.11].

CHi is important reductive radical in the reaction model. Ox (OH) is an important oxidative radical. We thought that concentrations of these radicals strongly depended on SRgas. Acording to figure 7, NOx was mainly reduced by CHi to form XN (NHi, HCN). The XN (NHi, HCN) formed NOx again. However, the remainder of XN reacted with NOx to form N_2. The main reductive radicals were CHi in this reaction mechanism, and the main oxidative radical was Ox (OH). Concentrations of these radicals were strongly influenced by the SRgas, and the gas-phase reaction between NOx, XN (NHi, HCN) and CHi seemed to reached the pseudo-steady state. Hence, NOx concentration strongly depended on SRgas.

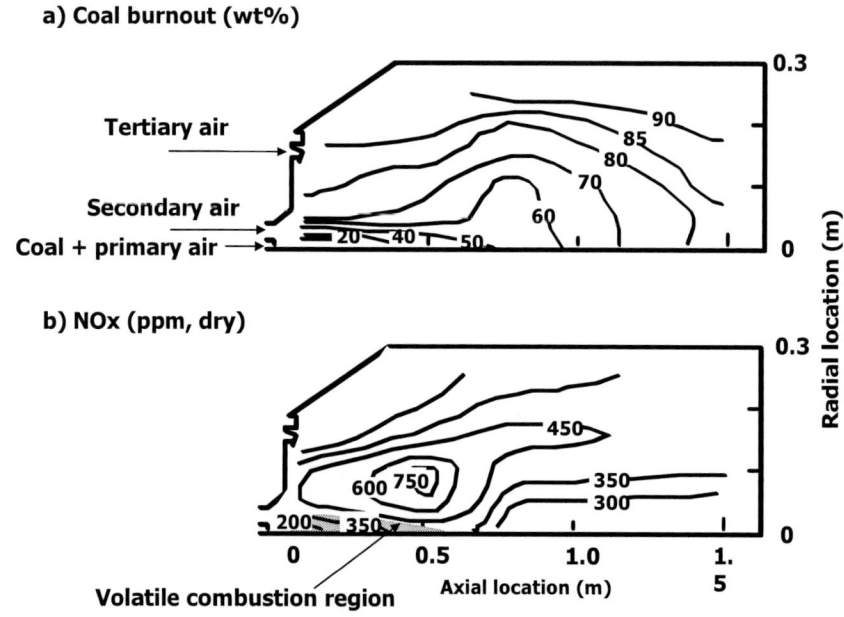

Figure 9. Examples obtained with the laboratory scale furnace of a) coal burnout and b) NOx distribution in flame produced by hv bituminous coal [ref.11].

An estimation method for Ox (OH) concentration, expressed by equation (6), has been shown previously [11]. The mole fraction of OH was derived from Bose and Wendt [15].

$$[OH] = 1.3 \times 10^4 \exp(13000/T) \times [OH_{eq}] \tag{6}$$

An example of the relation between SRgas and Ox (OH) is shown in figure 10.

We estimated CHi in the following way. CHi was mainly formed from the part of the fuel which was released from the solid phase to the gas phase by the pyrolysis, oxidation, and gasification reactions. We assumed that the amount of CHi formation increased with the rates of these reactions. Hydrogen was necessary for CHi formation. When hydrogen content in solid fuel increased, the CHi formation rate increased. It was easy to form CHi in the volatile combustion region because more hydrogen was contained in the volatile matter than in the char. Finally, we concluded that SRgas was the most important factor. We estimated CHi concentration by equation (7).

[CHi] = f (SRgas, burning temperature,

[summed rates of pyrolysis, oxidation and gasification reactions], hydrogen content in combustible solid)   (7)

We estimated CHi concentration under the burning conditions in the drop-tube furnace for a hv bituminous coal. The calculation method for combustion, heat transfer, and fluid dynamics was shown previously [7, 11]. Figure 11 shows the relation between SRgas and estimated CHi. For comparison, measured CH4 concentration is also shown in the figure. We considered that methane was one of the most similar and stable species to CHi. It is hard to measure radical species in pulverized coal flames. Our proposed method of estimation was judged to be proper because the relative concentration changes of estimated CHi were similar to those of measured methane. Estimated CHi concentration increased when burning time became short, even if SRgas was the same. Similar phenomena were observed for measured CH_4. This was because the char concentration increased when burning time became short. When the char concentration increased, the reaction rate of oxidation and gasification increased. Then the release rate of C and H from solid to gas phase increased. We judged the proposed NOx reaction model was proper for the most part.

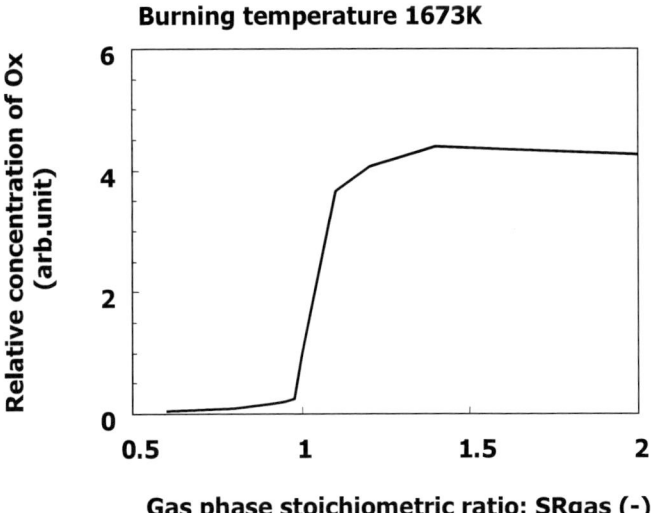

Figure 10. An example of the relationship between SRgas and estimated Ox(OH) concentration for hv bituminous coal. Flame temperature was 1673K.

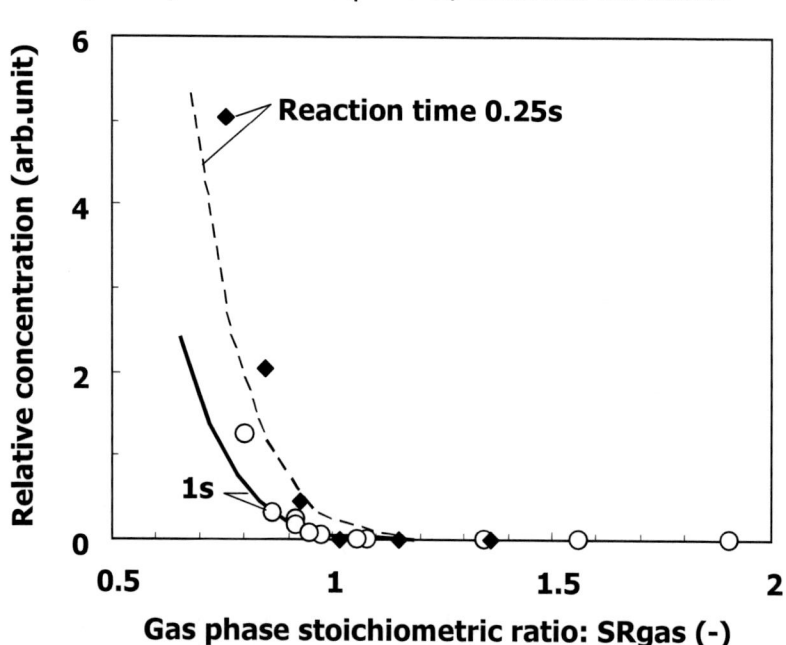

Figure 11. An example of the relationship between SRgas and estimated CHi concentration for hv bituminous coal. Flame temperature was 1573K.

Experimental and calculated results are compared in figure 12. Agreement was good for both air combustion and oxy-fuel combustion.

We have also developed a set of CFD programs based on such basic experimental data [1, 16]. The set of programs can calculate chemical reaction, fluid dynamics, and heat transfer for various kinds of solid fuels. Figure 13 shows typical results of the CFD calculations for an actual power plant. This program can simulate the detailed structure of heat exchangers precisely in order to calculate detailed gas and steam temperature distributions [17]. By using LES, detailed fluid dynamics of the ignition region of the burner can be calculated [18]. In numerical analysis, the making of the mesh takes much time and even for an expert, mesh making for a complicated system such as a power plant system is difficult. We developed a tool which made the mesh automatically based on data of furnace shape and operation conditions. We can input the data by using the GUI on a Web browser. This system reduces work time for mesh generation to 1/10 and allows performance evaluation of an actual power plant.

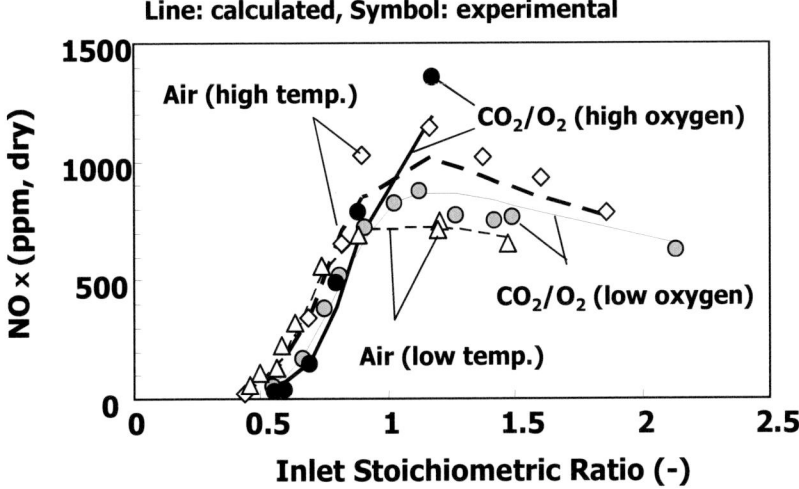

Figure 12. Verification of the NOx reaction model by the drop-tube furnace experiments.

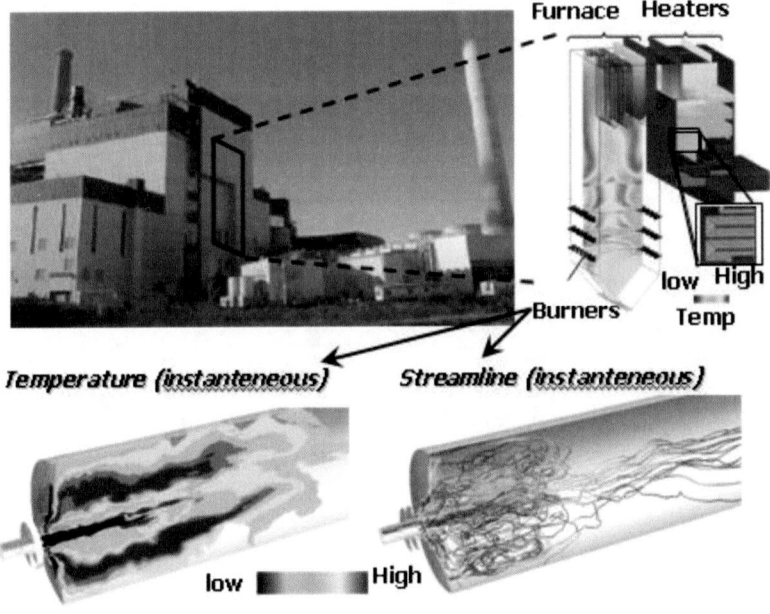

Figure 13. Example CFD calculations for an actual power plant (pulverized coal fired boiler system) [ref.1].

An example of verification for an actual boiler is shown in figure 14. Electric output of the boiler was 1000 MW; furnace structure, size, and driving conditions are shown in the figure and were from the literature [19]. Other details of the boiler structure have been shown in the literature [19-21]. We did calculations under several operating stoichiometric ratios.

Calculated NOx emission and the unburned carbon content in fly ash agreed for the most part with measured values. Usually, it is difficult to improve the NOx emission performance and lower the amount of unburned carbon in fly ash, simultaneously. For example, when we reduced NOx emission, the amount of unburned carbon in fly ash easily increased. The calculated results agreed well for these features of actual results.

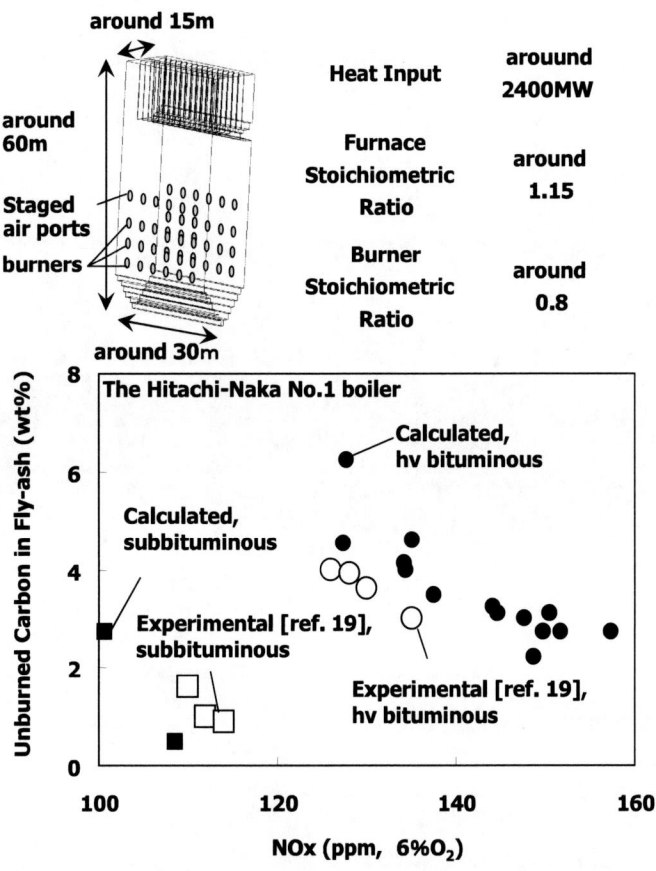

Figure 14. Validation of the proposed model for a commercial scale boiler.

2.3. CALCULATION FOR OXY-FUEL COMBUSTION

Example calculated results for oxy-fuel combustion are shown in figures 15 and 16. Figure 15 compares differences in NOx emission and unburned carbon content in fly ash between air combustion and oxy-fuel combustion. Because the performance results of an actual boiler have not been published, we compared the calculated results with experimental results in a pilot scale furnace [12]. The average oxygen concentration in combustion supporting gas was 27-30vol% for oxy-fuel combustion. NOx emissions for oxy-fuel combustion were reduced to 1/4 to 1/3 of those for air combustion. The unburned carbon content in fly ash also decreased for oxy-fuel combustion. The calculated results reproduced the features of the experimental ones. Although we had only a small number of verifications, we felt the reaction model could be reasonably applied for oxy-fuel combustion.

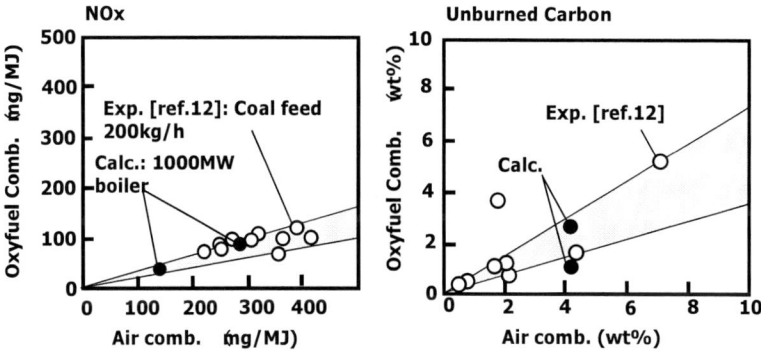

Figure15. Comparison of combustion performances between air combustion and oxy-fuel combustion. For oxy-fuel combustion, the average oxygen concentration in combustion supporting gas was 27-30vol%.

Figure 16 shows calculated NOx concentration distributions in air combustion and in oxygen combustion flames. NOx increased rapidly near the burner, then, NOx reduced downstream in the fuel-rich region of the burner zone. This characteristic was similarly examined for air combustion and oxygen combustion. NOx was regenerated again after the staged gas (air or CO_2/O_2) mixed with combustion gas in the burner zone. The amount of the regenerated NOx for oxy-fuel combustion was smaller than that for air combustion. A fuel-lean flame was formed right after the staged gas mixed with the combustion gas in the burner zone. For air-burning bolers, thermal

NOx is formed in this region. On the other hand, thermal NOx formation for oxy-fuel combustion was very small, because only a small amount of N_2 was contained in the combustion gas. The amount of regenerated NOx was small. Figure 16 also shows NOx and CO emissions (amounts of NOx and CO exhausted from the system), and unburned carbon content in fly ash. NOx emission decreased for oxy-fuel combustion, as reported in a past investigation [22]. CO emission and unburned carbon content in fly ash also decreased. One of the main reasons for this was that the average oxygen concentration in combustion supporting gas was higher than 21vol%.

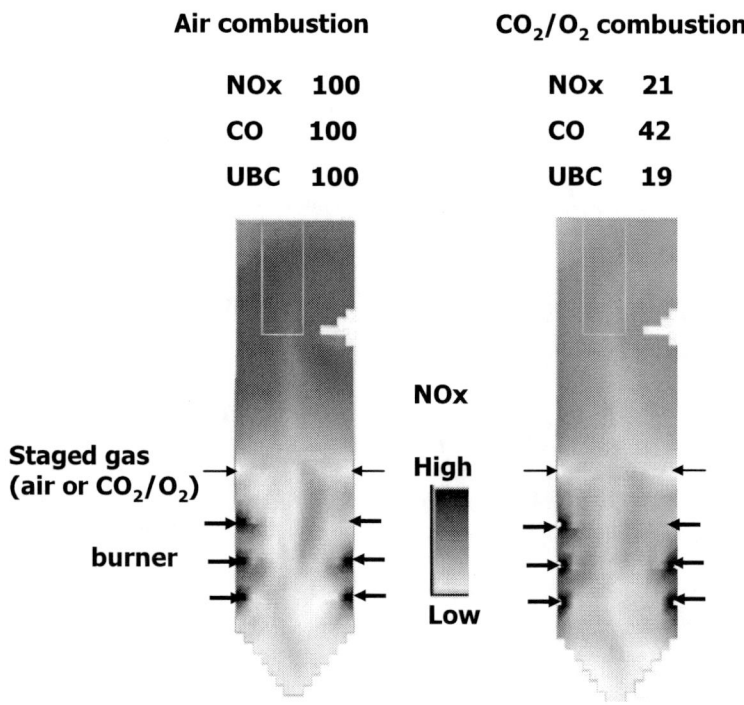

Figure 16. Calculated NOx distribution in air combustion flame and oxy-fuel combustion flame. Amounts of NOx and CO exhausted from the system and unburned carbon in fly ash at the exit of the furnace are also shown. The shape of the furnace and the operating conditions were the same as those shown in Fig.14. UBC: unburned carbon content.

Chapter 3

COAL IGNITION REACTION MODEL

3.1. COAL IGNITION STUDY
FOR OXY-FUEL COMBUSTION

The ignition property is one of the most important combustion performance parameters for engineering design of combustion systems. The number of coal ignition studies for oxy-fuel combustion has been increasing [23-26]; however, there are still only a few studies compared with the number of ignition studies for air combustion. Further examinations are necessary for future development. Arias et al. [23] studied ignition temperatures of coal and biomass under various oxygen concentrations of CO_2/O_2 atmospheres. Man et al. [24] studied ignition conditions for low volatile coals. Shaddix and Molina [25] studied pyrolysis and ignition delay time for single coal particles. Suda et al. [26] studied flame propagation velocities in the microgravity condition and measured flame propagation velocities under various conditions of combustion supporting gas. Flame propagation velocity is one of the most important ignition performance parameters. Lean flammability limit is also a very important ignition performance parameter to design combustion systems. However, the lean flammability limits of pulverized coals for oxy-fuel combustion have not been reported.

We developed a laser ignition experiment [27, 28] to improve the flame stability of a pulverized coal burner. In the experiment, pulverized coal was suspended in an upward flow, and then ignited by a pulsed laser. The advantage of the experiment was that similar ignition phenomena to those of an actual combustor could be observed by using a very little sample amount.

Both flame propagation velocity and lean flammability limit could be obtained.

We measured both the flame propagation velocities and lean flammability limits of N_2/O_2 atmospheres, for various kind of coals and particle diameters. We analyzed results of Suda et al. [26] and our own results, and developed a model for estimating lean flammability limit for oxy-fuel combustion.

Furthermore, we discussed a method to predict blow-off limits for actual combustion systems, based on fundamental experimental results. The quantity of heat loss from a flame to a furnace wall varies with the scale of the experimental equipment, i.e., laboratory-scale and pilot-scale furnaces and actual boilers [29]. In this study, we also considered a method to correct for the influence of heat loss.

3.2. EXPERIMENTAL PROCEDURE

Figure 17 shows the laser ignition equipment [27, 28]. Uniformly sized pulverized coal particles were suspended in a laminar upward flow and rapidly heated (heating rate, 10^5-10^6 K/s) by a single-pulsed YAG laser (pulse duration, 150μs; maximum laser energy, 1J/pulse; energy flux, about 10^8w/m^2; wavelength, 1064nm). Velocity of the upward flow was controlled according to the particle diameter. The heated pulverized coal particles were burned in the quartz test section (50mm square cross section). A He-Ne continuous sheet laser (sheet width, 3 x 10mm; energy flux, around 2×10^{-2} w/m^2) was irradiated horizontally at the ignition point. The particle concentration was measured from the intensity of particle scattering by the He-Ne laser.

Another continuous laser (copper vapor laser; maximum laser power, 15W; wavelength, 488 or 515nm) was used to determine the effect of radiant heat loss on the ignition characteristics. The radiant heat loss was larger for the basic equipment because the quartz wall of the cross-sectional area was kept at room temperature. The effect of radiant heat loss was examined by varying the energy flux of the copper vapor laser. The laser beam was supplied through an optical fiber. The beam diameter was controlled using a collimating lens and concave mirror and was usually 15mm around the ignition point. Emissions from the igniting/burning particles and scattering particles were detected using three photomultiplier tubes (PMTs) and the events were concurrently recorded by high-speed video (1000 frames/s; shutter speed, 10μs).

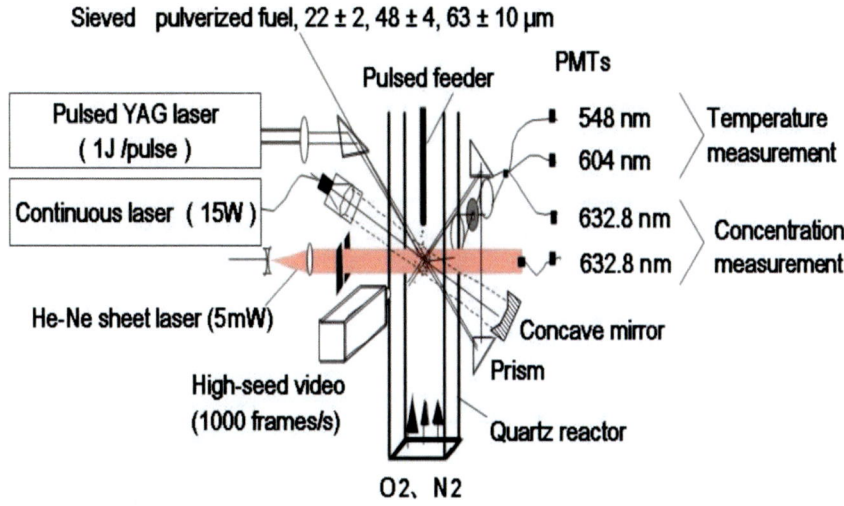

Figure 17. Schematic drawing of the laser ignition equipment [ref.28].

3.3. EXPERIMENTAL RESULTS
FOR N_2/O_2 COMBUSTION

The relation between the flame propagation velocity and particle concentration is shown in figure 18. The particle concentration was expressed as volatile matter concentration. Volatile matter concentration is the particle concentration multiplied by the volatile content of ultimate analysis.

The flame propagation velocity was defined as the increasing rate of flame diameter when the flame was approximated as a sphere [27]. We measured the flame propagation velocity of hv bituminous, lv bituminous, and anthracite coal samples. When volatile matter concentration was lowered to 0.02-0.03kg/m^3, flame propagation velocity of all coals was just beginning. The concentration of volatile matter strongly influenced the lean limit of flame propagation. When the volatile content increased, the flame propagation velocity became large. The maximum flame propagation velocity depended on the volatile content. When comparing coal types with the same volatile matter concentration, the flame propagation velocity increased in the order: anthracite < lv bituminous < hv bituminous. The volatile content was an important factor influencing flame propagation velocity.

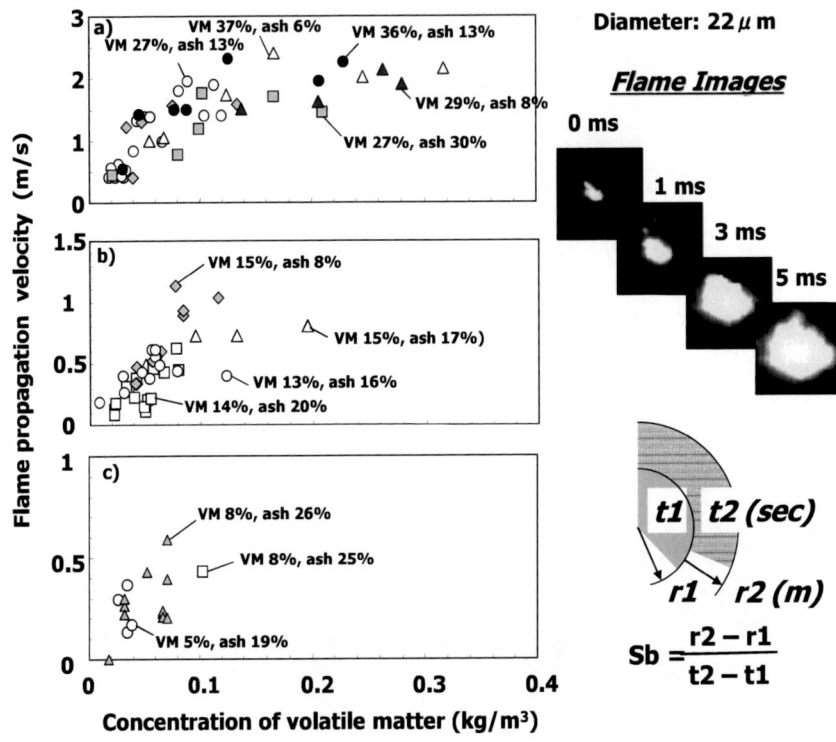

Figure 18. Relation between the concentration of volatile matter and the flame propagation velocity. Oxygen concentration was 100% and particle diameter of each coal type was 22±2μm [ref.28].

The relation between the volatile content and the lean flammability limit is shown in figure 19. The particle concentration at which the flame propagation probability becomes zero is defined as the lean flammability limit [27]. The flame propagation probability was calculated as shown by equation (8) [27].

Flame propagation probability
= number of experiments flame propagation was observed in/
number of experiments where ignition was observed   (8)

The calculations showed lean flammability limit was also strongly influenced by volatile content of coal.

We analyzed the relation between flame propagation velocities and lean flammability limits for various experimental conditions. The reciprocal of the lean flammability limit was almost proportional to maximum flame propagation velocity. Figure 20 shows the relation between the reciprocal of the lean flammability limit and flame propagation velocitiy.

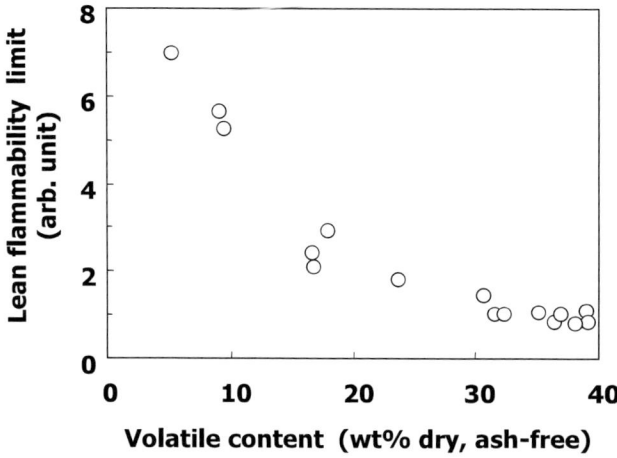

Figure 19. Relation between the volatile content (dry, ash-free basis) and the lean flammability limit. Oxygen concentration was 100% and particle diameter of each coal type was 22±2μm [ref.28].

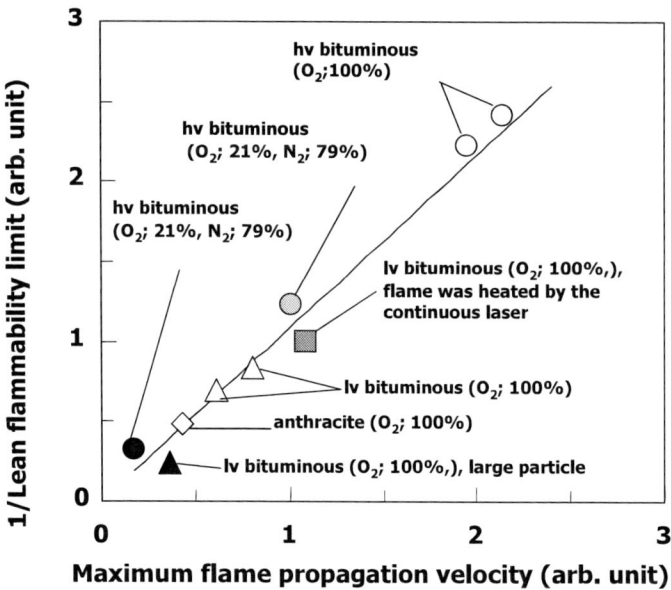

Figure 20. Relation between maximum flame propagation velocity and lean flammability limit.

3.4. LEAN FLAMMABILITY LIMIT
FOR CO_2/O_2 COMBUSTION

We measured flame propagation velocity in the N_2/O_2 atmosphere under various experiment conditions. Fujita et al. [30] suggested an experimental equation to express effect of O_2 concentration on flame propagation velocity, based on experimental results obtained by using the microgravity method. By using the same method, Suda et al. [26] measured flame propagation velocities in N_2/O_2, Ar/O_2, and CO_2/O_2 atmospheres. These experimental results are shown in figure 21. Fujita et al. had different coal property and particle size than our experiment had and Suda et al. had different particle size. The experimental results of Fujita et al. and Suda et al. are shown as corrected numerical values for our coal property (hv bituminous coal; volatile content, 27% dry basis) and particle size (diameter, $22 \pm 2 \mu$ m). For the N_2/O_2 atmosphere, the three sets of experimental results were almost in accord. Suda et al. only measured the flame propagation velocity with the CO_2/O_2 atmosphere. The fluctuation of the experimental results of Suda et al. is shown in figure 21. When the oxygen concentration was the same, the flame propagation velocities with the CO_2/O_2 atmosphere were smaller than those with the N_2/O_2 atmosphere.

Lean flammability limit for oxy-fuel combustion could be predicted by using lean flammability limits with N_2/O_2 atmosphere (figure 19), flame propagation velocities with N_2/O_2 and CO_2/O_2 atmospheres (figure 21), and the relation between lean flammability limit and flame propagation velocity (figure 20). Examples of predicted and experimental results are shown in figure 22. When the oxygen concentration was the same, the lean flammability limits for the hv bituminous coal with the CO_2/O_2 atmosphere were larger than those with the N_2/O_2 atmosphere. When the oxygen concentration was around 30vol% for oxy-fuel combustion, the lean flammability limit was the same as that for air combustion. When oxygen concentration was decreased, the lean flammability limit suddenly rose, and ignition became difficult. For the oxy-fuel combustion, the oxygen concentration at which the lean flammability limit suddenly increased was higher than that for the air combustion. When oxygen concentration was decreased to about 20vol%, ignition became extremely difficult for oxy-fuel combustion.

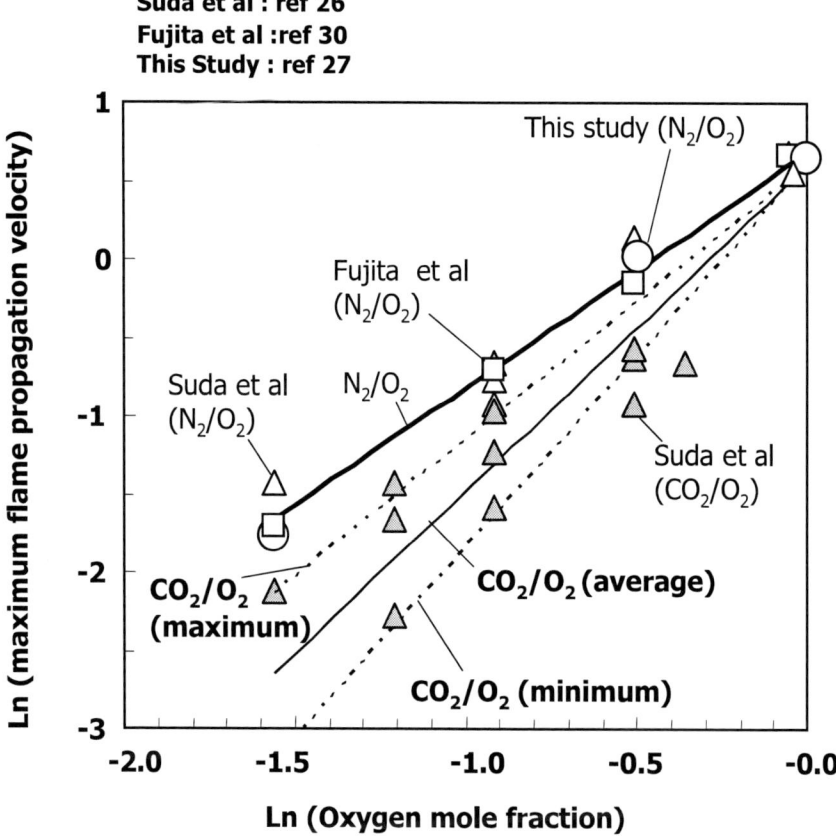

Figure 21. Effect of oxygen concentration on maximum flame propagation velocity for hv bituminous coals, particle diameter was $22\pm2\,\mu$ m; $\bigcirc$, $\triangle$, $\square$, obtained with N_2/O_2 combustion; $\blacktriangle$, obtained with CO_2/O_2 combustion; $\triangle$, $\blacktriangle$, the numerical values were corrected for particle diameter ($58\,\mu$ m $\rightarrow$ $22\,\mu$ m); $\square$, the numerical values were corrected for coal properties (subbituminous coal $\rightarrow$ hv bituminous coal).

Usually, coal is transported by carrier gas for pulverized coal fired boilers. The exhaust gas of combustion is usually used as carrier gas for oxy-fuel combustion systems. Oxygen concentration of the exhaust gas was too low to satisfy the ignition condition and addition of oxygen from the burner neighborhood would be needed to satisfy it.

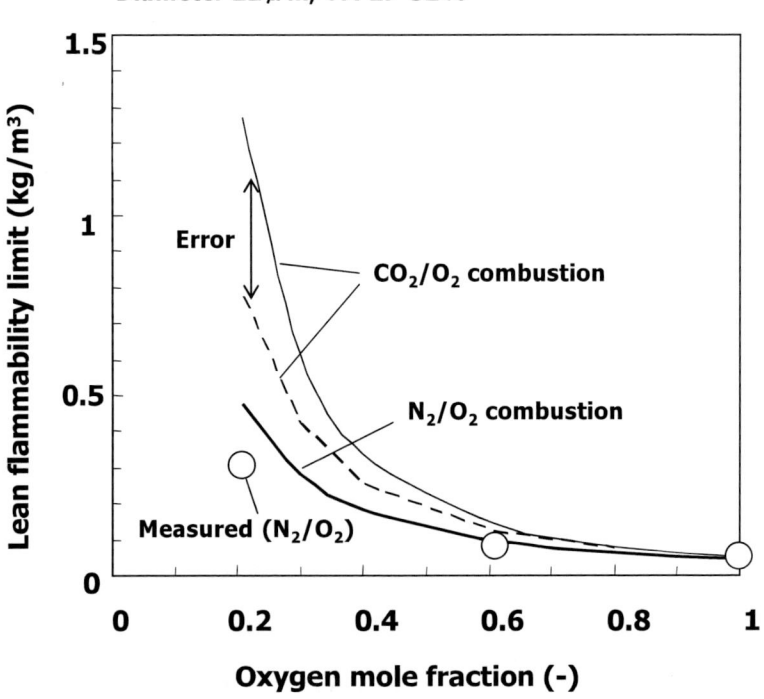

Figure 22. Effect of oxygen concentration on lean flammability limit for hv bituminous coal. The surrounding temperature (wall temperature) was 298K. Symbols, experimental; lines, prediction.

3.5. ESTIMATION OF BLOW-OFF LIMIT FOR LARGE-SCALE FURNACES

A pulverized coal mixture for which concentration is lower than the lean flammability limit cannot burn in an actual boiler or in an experimental device. But the heat loss rate from a flame to the surroundings (such as furnace wall) differs with the scale of the equipment. Thus, lean flammability limit (lean blow-off limit) for equipment varies with the heat loss rate. We looked next at the effect of heat loss rate on lean flammability limit (lean blow-off limit).

Figure 23 shows a model for flame stabilization of pulverized coal burners. Generally, pulverized coal and carrier gas (primary air) are supplied from the center of the burner. The secondary air is supplied around the primary air. In order to obtain a stable flame, it is necessary to raise pulverized coal concentration in the carrier gas (primary air) above the lean flammability limit. A recirculation region of the high temperature burning gas is formed between the flows of the primary air and secondary air by the effect of the flame stabilizer. The burning gas of the recirculation region heats the pulverized coal particles in the primary air and these particles ignite and burn. When the coal concentration is higher than the lean flammability limit, the burning coal particles heat and ignites other nearby coal particles. The flame moves from the coals directly heated by the gas of the recirculation region to the surrounding coal particles. This phenomenon is considered flame propagation. During the process of flame propagation, volatile matter in the coal burns and releases heat of combustion. This heat flows in the recirculation region and is fed back upstream. The flame is stabilized through repetition of these processes.

Here, Q is defined as the heat of combustion which flows in the recirculation region. Part of the heat Q that flows in the recirculation region radiates to the furnace wall. The remaining heat Q_{in} flows with the primary air from the recirculation region. First, we consider Q. We assume that Q is generated from combustion of volatile matter and is almost proportional to the amount of combustion for the volatile matter included in primary air. Q is expressed in equation (9).

$$Q = aQ_{vm}C \tag{9}$$

Here, a is a constant and Q_{vm} is volatile content of coal as the calorific value. C is the amount of fuel supply and usually it is proportional to coal concentration in the primary air. In the actual furnace, some part of Q is lost by radiant heat transfer to the wall. When we assume that the fraction of heat loss by radiation is $(1 - \alpha)$, the amount of radiant heat loss Q_{rad} is expressed as

$$Q_{rad} = (1-\alpha)Q \tag{10}$$

The difference in the amount of heat Q that is generated and the amount of radiant heat loss Q_{rad} becomes Q_{in} which flows in the primary air from the recirculation region. Q_{in} can be expressed as equation (11).

$$Q_{in} = \alpha Q = Q - Q_{rad} \tag{11}$$

Q_{min} is defined as the necessary minimum heat that should flow in the primary air to form a stable flame. The necessary condition to form a stable flame is expressed by equation (12).

$$Q_{in} \gtreqqless Q_{min} = \alpha \, a Q_{vm} C_{min} \qquad (12)$$

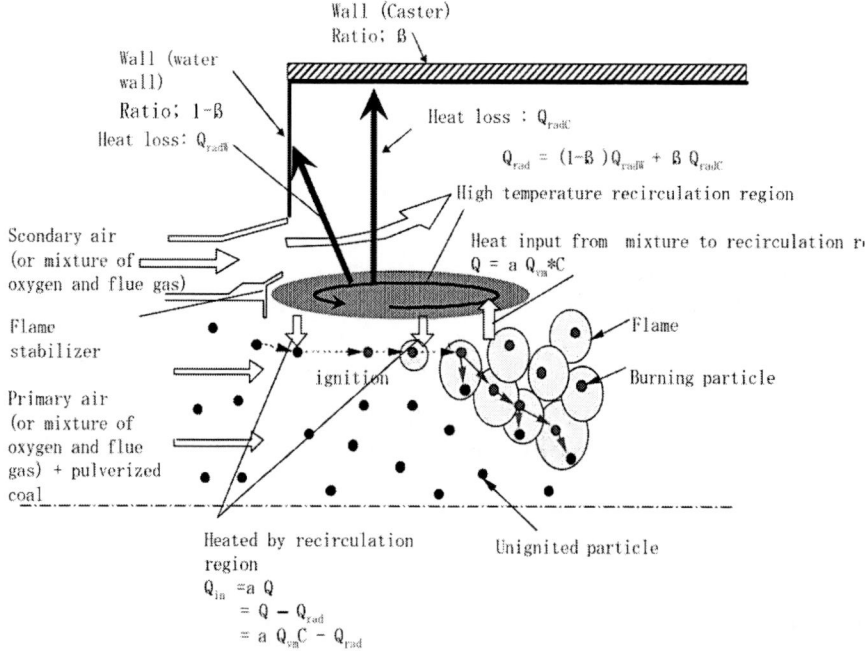

Figure 23. A model of flame stabilization near burners.

C_{min} is the minimum fuel supply. C_{min} is proportional to the lean flammability limit when the primary air supply is constant. Equation (13) can be obtained from (12).

$$1/C_{min} = \alpha \, (a/Q_{min}) \, Q_{vm} \qquad (13)$$

According to equation (13), the reciprocal of the minimum fuel supply is proportional to Q_{vm} (volatile content of coal as the calorific value, if the heat loss $(1-\alpha)$ is the same ($\fallingdotseq$ the burner and furnace are the same). As for the furnace shown in figure 23, a part of the furnace wall is covered with a caster, and the remainder is covered with a water wall. The heat loss is small for the caster-covered area, because wall temperature is high. The heat loss is large for the water wall part, because the wall temperature is low. We define the heat loss to the caster as Q_{radC} and the heat loss to the water wall as Q_{radW}. Judged from the viewpoint of the burner nozzle, the fraction of the area covered with the caster is defined as β. The area covered with the water wall is $1-\beta$. Total heat loss (Q_{rad}) can be expressed by equation (14).

$$Q_{rad} = (1-\beta)Q_{radW} + \beta Q_{radC} \qquad (14)$$

Here, if Q_{radW} is expressed as equation (15).

$$Q_{radW} = bQ_{radC} \ (b>1) \qquad (15)$$

Equation (14) is

$$Q_{rad} = Q_{radW} + \beta (b-1) Q_{radW}. \qquad (16)$$

When equation (14) is substituted for equation (16), Q_{in} can be expressed by equations (17) and (18).

$$Q_{in} = Q - Q_{radW} + \beta(1-b)Q_{radW} = \alpha Q \qquad (17)$$

$$\alpha = 1 - Q_{radW}/Q + \beta (1-b)Qr_{adW}/Q \qquad (18)$$

When equation (13) is substituted for equation (18), equation (19) is obtained.

$$1/C_{min} = (a/Q_{min}) Q_{vm} - (Q_{radW}/Q) (a/Q_{min}) Q_{vm}$$
$$+ \beta (1-b)(Q_{radW}/Q) (a/Q_{min}) Q_{vm} \qquad (19)$$

If the same burner and furnace are used and only the area fraction covered with a caster is varied, then only β in equation (19) changes. A straight line relation is obtained between the reciprocal of the minimum fuel supply ($1/C_{min}$) and the fraction of the area covered with the caster (β). The effect of heat loss on the condition of flame stabilization (such as blow-off limit) can be estimated from the relation between the measured $1/C_{min}$ and β.

When small-sized equipment is used, the effect of heat loss also can be estimated by heating the flame from surroundings by using devices such as lasers. When heating rate from the surroundings is defined as Q_{radL} for the case of a laser use, total heat loss (Q_{rad}) can be expressed by equation (20).

$$Q_{rad} = Q_{radW} - Q_{radL} \tag{20}$$

When equation (17) is substituted for equation (20), equations (21) and (22) are obtained.

$$Q_{in} = Q - Q_{radW} + Q_{radL} = \alpha Q \tag{21}$$

$$\alpha = 1 - Q_{radW}/Q + Q_{radL}/Q \tag{22}$$

When equation (13) is substituted for equation (22), equation (23) is obtained.

$$1/C_{min} = (a/Q_{min})*Q_{vm} - (Q_{radW}/Q)(a/Q_{min})*Q_{vm}$$
$$+ Q_{radL}(1/Q)(a/Q_{min})*Q_{vm} \tag{23}$$

A straight line relationship is obtained between the reciprocal of the minimum fuel supply ($1/C_{min}$) and heating rate from the surroundings (Q_{radL}).

We used a continuous wave laser for an energy source. Figure 24 shows the relation between laser energy and lean flammability limit. A straight line relation was observed between the laser energy (radiant heat flux from the continuous wave laser) and the reciprocals of the lean flammability limit.

Kiga et al. [29] measured the blow-off limit for pulverized coal flames by using large-scale equipment. We examined the effect of heat loss on blow-off limit more by analyzing their experimental results. Structures of the experimental equipment are shown in figure 25 [29]. Part of the pilot-scale furnaces was covered with caster. Blow-off limits were measured in three pilot-scale furnaces and an actual boiler by using the same design of burners. Coal feed rate of the pilot-scale furnaces was 3.2t/h. Changing the area of the caster varied the heat loss rate. For the experiment with the actual boiler, blow-off limit was measured when two sets of burners were operated. All the wall of the actual boiler was covered with a water wall. Judging from one flame, there was one part surrounded by the water wall directly, and there was another part that between one flame and the water wall. We assumed that heat loss rate was reduced to the same level as the caster wall if there was another flame between one flame and the water wall. We assumed that this area was covered with caster. The value of β became 0.0625.

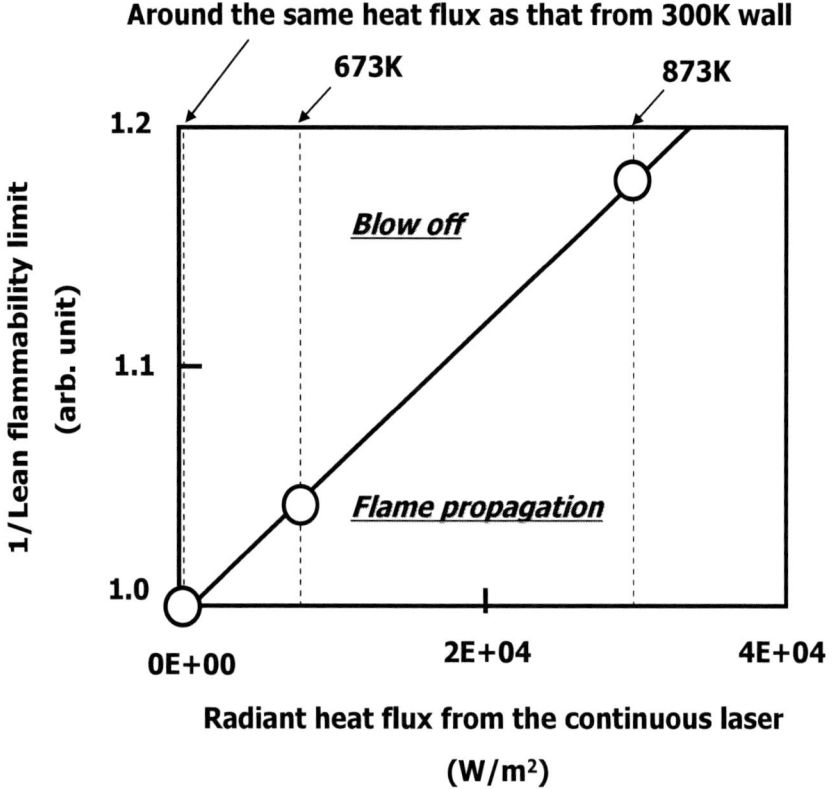

Figure 24. Effect of radiant heat flux from surroundings on the lean flammability limit for the laser ignition experiments.

The relation between β and the blow-off condition of the minimum fuel supply for stable combustion, $1/C_{min}$, shown in figure 26, was a straight line. The effect of the heat loss rate on blow-off limit could be estimated by using the proposed method. Blow-off limit for actual boilers could be predicted from experimental results for pilot-scale furnaces. Plural burners are usually operated for actual boilers. Blow-off limit and heat loss rate change depending on which burners are operated. In this case, the effect of heat loss rate can be estimated with the proposed method shown in figure 25.

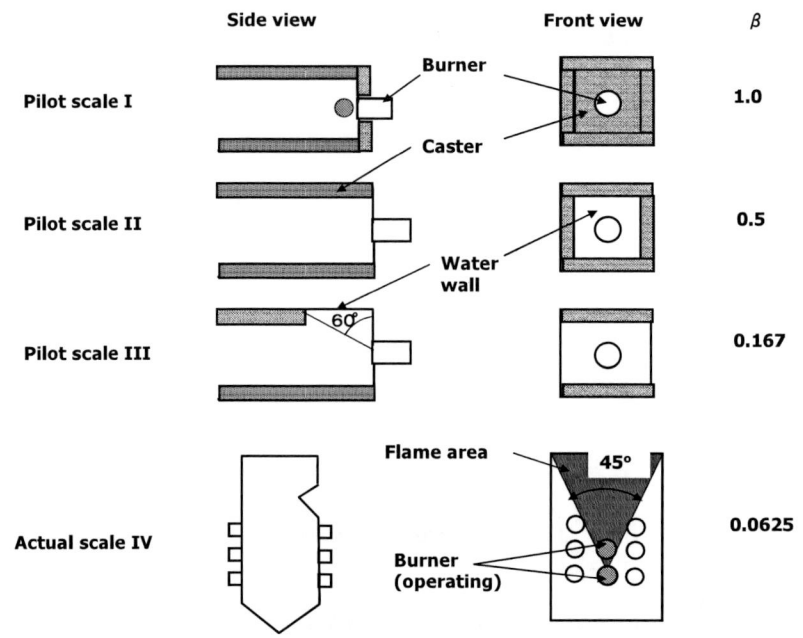

Figure 25. Structure of the furnace for blow-off experiment of large scale equipment [ref.29].

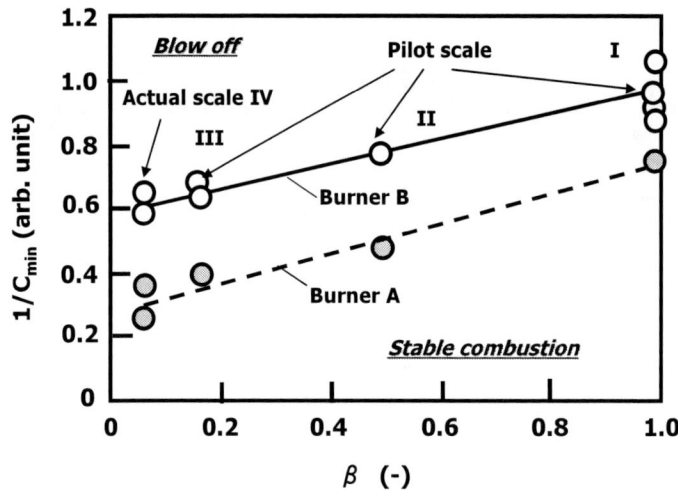

Figure 26. Effect of the β value (the ratio of an area covered with the caster judged from the viewpoint of the burner nozzle) on minimum fuel supply for stable combustion. The experimental data were obtained from ref.30.

According to the proposed model, the effect of coal properties on blow-off limit is expressed by equation (13) in which the reciprocal of the minimum fuel supply to stabilized flame (C_{min}) is proportional to volatile content of coals as the calorific value (Q_{vm}). The volatile content of the mass standard is roughly proportional to the volatile content of coals as the calorific value. Figure 27 shows the relation between volatile content of coal and C_{min}. A straight line relation was observed between volatile content of coal and C_{min} (or lean flammability limit). The relation shown by equation (13) was almost realized.

Using the proposed model and an experimental database, we calculated the lean flammability limit in a large-scale burner [28]. Figure 28 shows the relation between calculated and experimental blow-off limits. The calculation included the effect of heat loss, volatile content of coal, and particle diameter distribution. The calculated results agreed well with the experimental results.

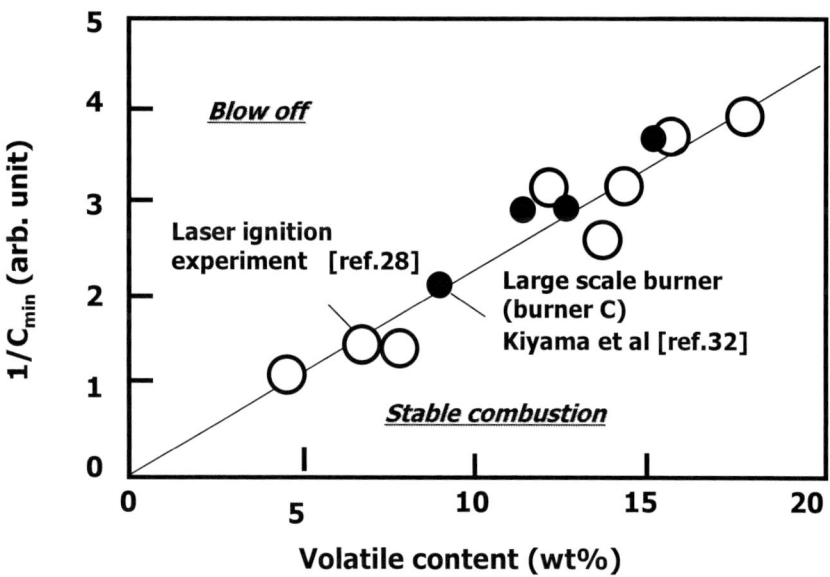

Figure 27. Effect of volatile content of coal on minimum fuel supply for stable combustion.

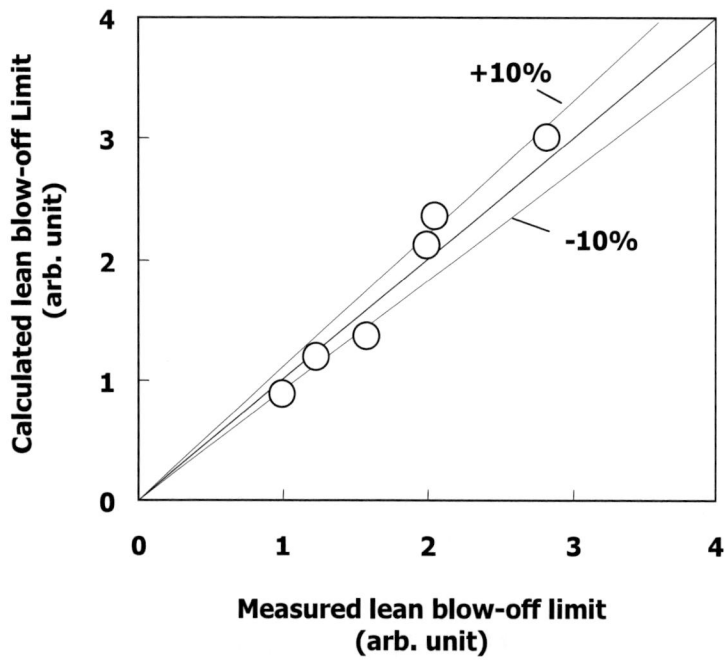

Figure 28. Relation between calculated and experimental blow-off limits.

Chapter 4

A CASE STUDY FOR OXY-FUEL COMBUSTION SYSTEMS

In this section, we introduce examples of numerical analyses for oxy-fuel combustion systems obtained by using the proposed models shown in Sections 2 and 3 and the CFD program shown in Section 2. In particular, we examined how to operate an oxygen combustion system when low-rank coal is used as fuel. Low rank coals, such as anthracite, are very hard to ignite when they are used for air combustion. Coal properties used for analyses are shown in table 1. Coal A is an example of an anthracite coal. Coal A is not used in pulverized coal filing boilers so much, because it is inferior in its firing and combustion performance. Coal B is an example of hv bituminous coal and it is widely used in pulverized coal fired boilers. We examined the system constitution of the oxygen combustion boiler by numerical analysis when these coals were used as fuels. Combustion performance parameters, such as ignition performance, heat absorption by the furnace wall, combustion efficiency (i.e. unburned carbon content in fly ash), and NOx emission were predicted for several furnace structures and operating conditions.

Table 1. Coal properties used for numerical analyses

Coal	HHV (MJ/kg)	Water (wt%)	VM (wt%, dry)	ash	C	H	O	N	S
					(wt%, dry ash-free)				
A	22.7	10	6.7	29	91.6	3.8	1.7	1.7	1.2
B	25.8	14	35	11.5	81.9	5.3	10.1	2.2	0.5

Ignition performance was examined at first. Figure 29 shows calculated relations between coal concentration and flame propagation velocity. The lines in the figure show flame propagation velocities in air atmosphere. Lean flammability limit is defined as the coal concentration when flame propagation velocity becomes zero. It is necessary to raise coal concentration in carrier gas (a mixture of primary air and coal) from the lean flammability limit in order to form a stable flame. The flame propagation velocity needs to be raised more for low-NOx combustion. The coal concentration should be raised as much as possible. But, the upper limit of the coal concentration is limited by the coal pulverizer used. In general, it is difficult to raise average coal concentration more than 1.5kg/m^3.

Case 1 in figure 29 showed flame propagation velocity when coal B was burned in air. This type coal is widely used in air combustion boilers. We judged that ignition and combustion performances would be good enough if the flame propagation velocity was the same as used for case 1.

No.	sumbol / line	Coal	Surounding gas	particle diameter	oxygen concentration
1	———	B	air	medium	21vol%
2	– – –	A	air	medium	21vol%
3	– · —	A	air	fine	21vol%
4	△	B	oxyfuel	medium	21vol%
5	○	B	oxyfuel	medium	around 30%
6	◉	A	oxyfuel	fine	Larger than 30vol%

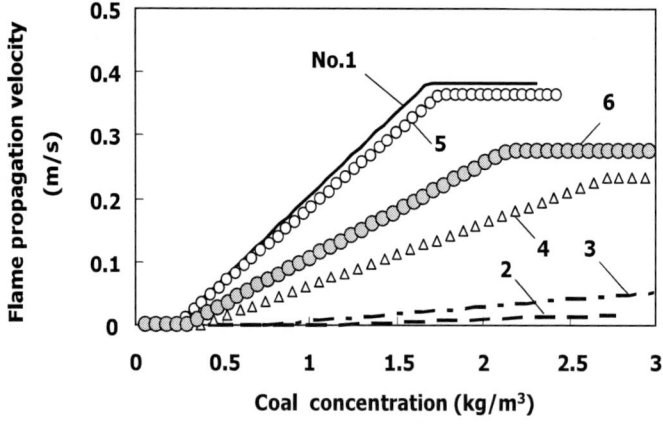

Figure 29. Calculated relations between particle concentration and flame propagation velocity. Temperature of the furnace wall was 1300K.

Case 2 in figure 29 showed flame propagation velocity when coal A was burned in air. Flame propagation velocity was too low to obtain a stable flame for pulverized coal combustion. Case 3 was the results when the particle diameter of coal A was fine. Usually, coal diameter must be decreased when ignition performance of the coal is inferior. The flame propagation velocity grew large when the coal diameter was decreased. Lean flammability limit became low. But the ignition performance was considerably inferior compared with coal B even if the coal diameter was fine. Increasing coal concentration is required if coal A is used for air combustion systems, but, it is difficult to secure sufficient ignition performance. The geometric symbols in figure 29 show results for cases 4 to 6 of different flame propagation velocities in CO_2/O_2 atmosphere. Case 4 gave the flame propagation velocity of coal B for oxy-fuel combustion when oxygen concentration was 21vol%. The flame propagation velocity was around 1/3 of its value for air combustion. It would be necessary to increase the oxygen concentration or coal concentration in carrier gas in order to obtain the same ignition performance as that of air combustion. Coal concentration should be about three times larger than that of air combustion if oxygen concentration was 21vol%. Case 5 shows flame propagation velocity of coal B when oxygen concentration was increased. Both flame propagation velocity and lean flammability limit were almost equal with those of air combustion by controlling oxygen concentration. In this case, ignition performance was equal with that of air combustion when oxygen concentration was around 30vol%. But the appropriate oxygen concentration varied with the coal property and diameter.

Case 6 showed flame propagation velocity of coal A for oxy-fuel com-bustion. Coal diameter was the same as that of Case 3. Oxygen concen-tration was more than 30vol%. For oxy-fuel combustion, oxygen concentration in combustion supporting gas can be changed easily. This is one of the advantages of oxy-fuel combustion. By combining this advantage with a fine grain coal, even if it was coal A (anthracite), good ignition performance that was near that of coal B (hv bituminous coal) was provided.

We examined the arrangement of gas ports (burners and gas ports for staged combustion) next. We also examined the distribution of oxygen and flue gas (combustion exhaust gas supplied by recirculation) supply to the gas ports. We wanted to obtain a uniform heat flux distribution to the furnace wall. Figure 30 shows the effect of the arrangement of the gas ports and the distribution of oxygen and flue gas supply on heat flux distribution to the furnace wall. Figure 30a shows heat flux distribution when the ratios of oxygen and flue gas supply to all gas ports were the same. Heat flux distribution in the burner part was high and the heat flux at the upper part of the furnace was low. If the heat flux distribution was highly irregular, it was easy to get an irregular temperature distribution in the steam generated from the furnace. In addition, the life of the furnace would be shortened if the heat flux distribution was very irregular.

Figure 30b shows calculation results when the ratio of oxygen and flue gas supply was varied at each gas port. The ratio of oxygen and flue gas supply to the burners decreased in order to decrease heat flux in the burner part. In this case, the lengthwise direction of the burner and the burner were also enlarged. The heat flux distribution varied with the ratio of oxygen and flue gas supply at each gas port. The adjustment of the irregularity of the heat flux distribution would be difficult for air combustion, because it is hard to change oxygen concentration. For the oxy-fuel combustion, it could get closer to a uniform heat flux distribution by a comparatively simple adjustment.

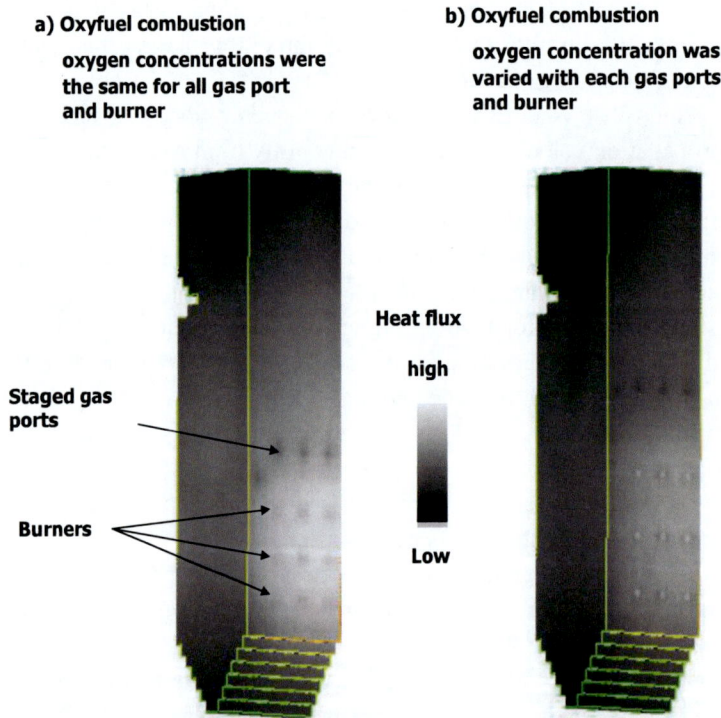

Figure 30. Heat flux distribution to the furnace wall obtained with different operating conditions.

Figure 31 shows schematic side views of the furnaces and example temperature distributions. Figure 31a is the shape of a usual boiler (furnace 1). This is widely used for air combustion. Staged combustion was used to reduce NOx emission. Six sets of burners were arranged in the furnace. All the wall of the furnace was a water wall. Temperature of the water wall was assumed as 673K for the calculation. We chose the most suitable position according to the coal properties. The position of the gas ports (burners and gas ports for staged combustion) in figure 31a was used when coal B was burnt in air. The furnace shown in figure 31b (furnace 2) was used when coal A was burnt in air in order to improve the ignition performance. A caster was installed in the wall of the burner part to reduce the heat loss to the wall from the flame. The flame temperature in furnace 2 became higher in comparison with that in furnace 1.

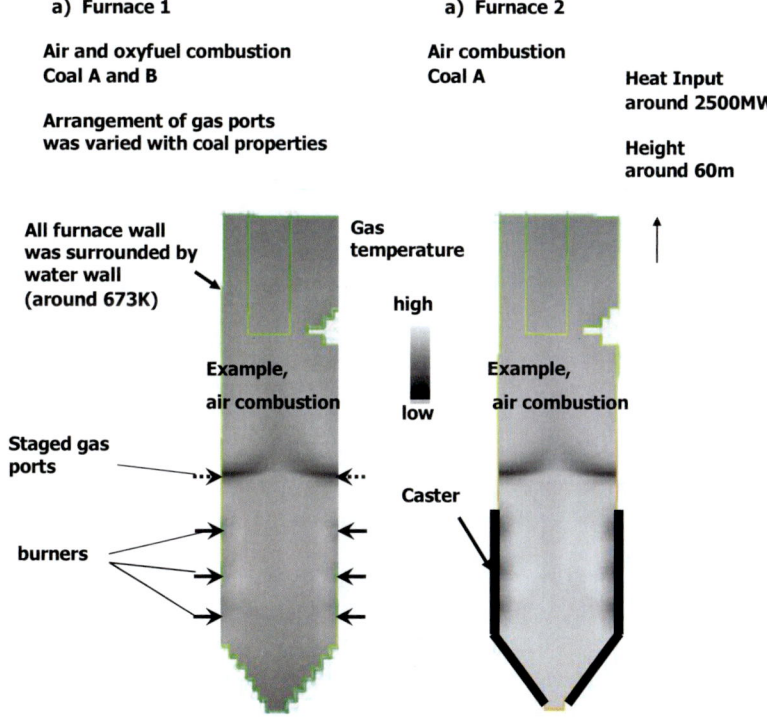

Figure 31. Schematic side views of the furnaces and examples of temperature distribution.

The flame propagation velocity was calculated by the proposed model shown in Section 3. Calculation conditions are shown in table 2. For coal A, particle concentration was increased and particle diameter was decreased. Oxygen concentration in the carrier gas was increased for oxy-fuel combustion. Wall temperature and β are also shown in the table. The value of β varied with furnace shape. The evaluation for β is explained schematically in figure 32 and was the same as that shown in figure 25 the blow-off experiment. In the burners of furnace 1, there were many parts facing the water wall. Lean flammability limit was calculated at first. The influence of the wall temperature on lean flammability limit was corrected for with the results of figure 24. Maximum flame propagation velocity was evaluated from the results of figure 20 (relation between lean flammability limit and maximum flame propagation velocity). The relation between flame propagation velocity and coal concentration was estimated from the results of figure 18.

Table 2. Operating conditions used for evaluation of flame propagation velocities

CASE No.	Coal	Diameter	Oxygen concentration	Combustion system	coal concentration in carrier gas	Furnace	β	Wall temperature (caster or flame)	Wall temperature (water wall)
1	B	under 200mesh; 81wt%	21vol %	air	1.0 kg/m3	1	0.188	1573K	673K
2	A	200mesh; 81wt% under	21 vol%	air	1.0 kg/m3	1	0.188	1573K	673K
3	A	200mesh; 91wt% under	21 vol %	air	1.5 kg/m3	2	1.00	1573K	673K
4	B	200mesh; 81wt% under	around 30 vol %	oxyfuel	1.0 kg/m3	1	0.188	1573K	673K
5	A	200mesh; 91wt% under	larger than 30 vol %	oxyfuel	1.5 kg/m3	1	0.188	1573K	673K

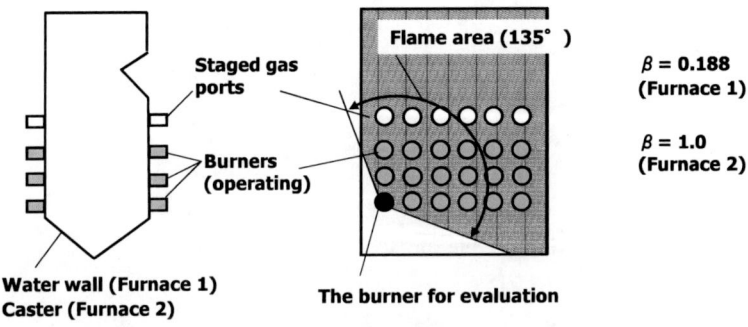

Figure 32. Evaluation of β value for the furnaces.

Figure 33 shows our calculated results. Cases 1, 2 and 3 were calculation results of air combustion. Cases 4 and 5 were results of oxy-fuel combustion. Flame propagation velocity, NOx emission, unburned carbon content in fly ash, and, heat absorption by the furnace wall are shown in the figure.

Flame propagation velocities were shown as the value that was corrected for the effect of heat loss rate for a burner installed at the position where heat loss rate was the largest as shown in figure 32. Case 1 showed the calculation results when coal B (hv bituminous) was burned in air. Ignition performance of other cases was evaluated against case 1 as the standard. Flame propagation velocities of coal B in air (cases 2 and 3) were inferior to case 1. The flame propagation velocity of case 2 (without caster) was particularly small. Stable combustion was difficult considering the load changes of the boiler. The flame propagation velocity of case 3 (with caster) was improved by around 70% compared to case 1. When coal A (anthracite) is used in an air combustion boiler, it is necessary to prevent heat loss to the furnace wall from the flame. Cases 4 (coal B) and 5 (coal A) were results of oxy-fuel combustion. Flame propagation velocities of both coals A and B rose by controlling oxygen concentration in the carrier gas (primary gas) adequately.

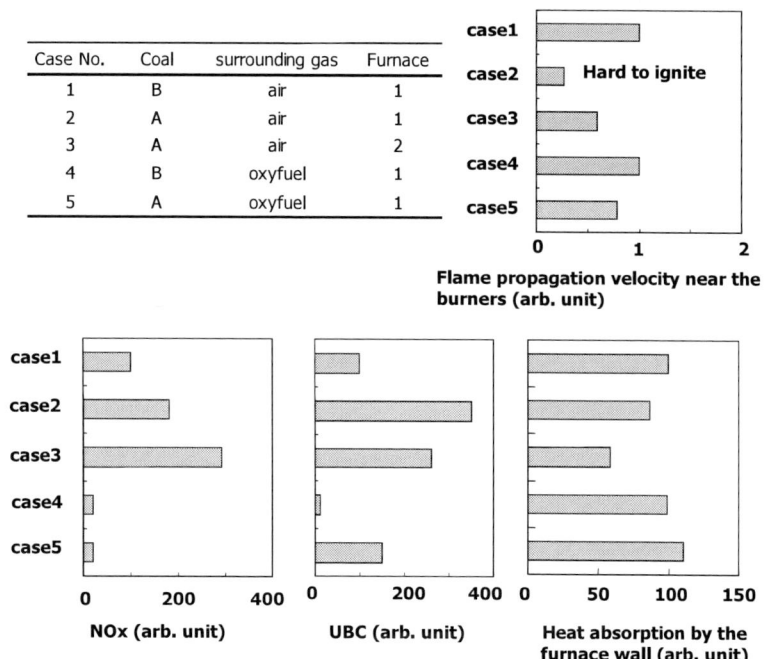

Figure 33. Calculated results of case studies.

We evaluated NOx emission, unburned carbon content in fly ash, and heat absorption by the furnace wall next. When coal A was burned in air, NOx emissions and unburned carbon in fly ash increased in comparison with case 1. The caster was installed in case 3 to improve the ignition performance. NOx emissions of case 3 rose very much because burning temperature became higher and the formation rate of thermal NOx increased. Heat absorption by the furnace wall was reduced because the caster was installed. When coal A was burned in air, it was difficult to obtain equal combustion performance to that of case 1. However, when coal A was used for oxy-fuel combustion (case 5), NOx emissions and unburned carbon in fly ash could be decreased significantly. The NOx emissions were close to those of case 4 (coal B for oxy-fuel combustion). Unburned carbon in fly ash was inferior to that of case 4, but was at the same level as case 1 (coal B for air combustion). Heat absorption by the wall was the same level as case 1. A high-efficiency and low-emission combustion system was possible for oxy-fuel combustion, even if the coal used was difficult to ignite and burn.

5. CONCLUSION

1) The NOx reaction mechanism was investigated for air combustion and oxy-fuel combustion. We proposed the gas phase stoichiometric ratio (SRgas) as a key index to evaluate NOx concentration in fuel-rich flames. The SRgas was defined as:

SRgas ≡ amount of fuel required for stoichiometry combustion
        /amount of gasified fuel

where the amount of gasified fuel was defined as the amount of fuel which had been released to the gas phase by pyrolysis, oxidation and gasification reactions. When SRgas<1.0, NOx concentration was strongly influenced by the value of SRgas, but it was hardly influenced by burning conditions. A NOx reaction model was developed by using the index. The NOx concentration for a drop-tube furnace and NOx emissions of a commercial scale boiler could also be reproduced by using this model.

2) Flame propagation velocity and lean flammability limit of pulverized coal were examined under various coal properties and burning conditions. We developed an empirical formula to express the relation between flame propagation velocities and lean flammability limit. With this formula, we predicted lean flammability limit of pulverized coal in CO_2/O_2 atmosphere. We examined the effect of radiant heat loss to the furnace wall from the flame on lean flammability limit by analyzing experimental data obtained with small- and large-scale equipment. A model to estimate blow-off limit was

developed based on the results. The calculated blow-off limits agreed with experimental results obtained with the large-scale equipment.

3) We carried out a case study of the oxygen combustion boiler system using the proposed ignition model and CFD program. The coal reaction model of the CFD program was developed based on the present fundamental studies. It is easy to control oxygen concentration for oxy-fuel combustion. This is one of the advantages of oxy-fuel combustion. By using this advantage, a high-efficiency and low-emission combustion system could be realized, even if the coal used was difficult to ignite and burn.

REFERENCES

[1] O. Ito, K. Chino, E. Satito, S. Marushima, C. Bergins and S. Wu, "CO_2 reduction technology for thermal power plant systems." *Hitachi Review,* 57(5) (2008) 166-173.

[2] K. D. Tigges, F. Klauke, C. Bergins, K. Busekrus, J. Niesbach, M. Ehmann, C. Kuhr, F. Hoffmeister, B. Vollmer, T. Buddenberg, S. Wu and A. Kukoski, "Conversion of existing coal–fired power plants to oxy-fuel combustion: case study with experimental results and CFD-simulations." *Energy Procedia,* 1 (2009) 549-556.

[3] R. Gupta S. Khare, T. Wall, K. Eriksson, D. Lundstrom, J. Eriksson and C. Spero., "Adaption of gas emissivity models for CFD based radiative transfer in large air-fired and oxy-fired furnaces." *The Proceedings of the 31st International Technical Conference on Coal Utilization & Fuel Systems,* May 21–26, FL, USA, (2006)

[4] K. Yamamoto, T. Fukuchi, M. Chaki, Y. Shimogori and J. Matsuda, "Development of computer program for combustion analysis in pulverized coal fired boilers." *Hitachi Review,* 49(3) (2000) 4976-4980. Available at: http://www.hitachi.co.jp/Sp/TJ-e/index.html.

[5] A. Williams, M. Pourkashanian, P. Bysh and J. Norman, "Modeling of coal combustion in low-NOx p.f. flames." *Fuel,* 73 (1994) 1006-1019.

[6] M. Xu, J. L.T. Azevedo and M. G. Carvalho, "Modeling of the combustion process and NOx emission in a utility boiler.", *Fuel,* 79 (2000) 1611-1619.

[7] K. Yamamoto, M. Taniguchi, H. Kobayashi, T. Sakata and K. Kudo, "Validation of coal combustion model by using experimental data of utility boilers", *JSME International Journal Series* B, 48 (2005) 571-578.

[8] T. Le Bris, F. Cadavid, S. Caillat, S. Pietrzyk, B. Blondin and B. Baudoin, "Coal combustion modeling of large power plant, for NOx abatement." *Fuel.* 86 (2007) 2213-2220.

[9] L. I. Diez, C. Cortés and J. Pallarés, "Numerical investigation of NOx emission from a tangentially-fired utility boiler under conventional and overfire air operation." *Fuel,* 87 (2008) 1259-1269.

[10] S. Phillips, N. Shinotsuka, K. Yamamoto and Y. Fukuda, "Application of high steam temperature countermeasures in high sulfur coal-fired boilers." Available at: http://www.hitachi. powersystems.us/supportingdocs/forbus/hpsa/technical_papers/EP2003B .pdf.

[11] M. Taniguchi, K. Yamamoto, H. Kobayashi and K. Kiyama, "A reduced NOx reaction model for pulverized coal combustion under Fuel rich conditions." *Fuel.* 81 (2002) 363-371.

[12] T. Yamada, "Current status of the Callide oxy-fuel Demonstration Project." In: *Clean Coal Day in Japan.* (2006) L7-1-20

[13] C. J. Tullin, S. Goel, A. Morihara, A. F. Sarofim and J. M. Beer, "Nitrogen oxide (NO and N_2O) formation for coal combustion in a fluidized bed: effect of carbon conversion and bed temperature." *Energy and Fuels,* 7 (1993) 796-802.

[14] S. Azuhata, K. Narato, H. Kobayashi, N. Arashi, S. Morita and T. Masai, "A study of gas composition profiles for low NOx pulverized coal combustion and burner scale-up." In: Twenty-First Symp. (Int.) on Combustion, Pittsburgh: The Combustion Institute, (1986) 1199-1206.

[15] A. C. Bose and J. O. L. Wendt, "Pulverized coal combustion: Fuel nitrogen mechanisms in the rich post-flame." In: Twenty-Second Symp. (Int.) on Combustion, Pittsburgh: The Combustion Institute, (1988) 1127-1134.

[16] K. Iwashige, H. Hamatake, M. Aoyama, K. Amamoto and H. Urai, "Numerical simulation for electric power plant systems." *Hitachi Hyoron,* (2008), 90(11) 66-71, in Japanese.

[17] M. Handa, K. Yamamoto, A. Yamada, K. Yamamoto and Y. Shimogori, "Nemerical method for three-dimensional analysis of shell– and tube– type of large scale heat exchangers under high temperature circumstances." *Advances in Computational Heat Transfer,* (2008) CHT-08.

[18] K. Yamamoto, T. Okazaki, T. Murota and M. Taniguchi, "Large eddy simulation of pulverized coal combustion." *Thermal Science and Engineering,* 15 (2007) 233-240, in Japanese.

[19] M. Rechardson, Y. Shimogori and Y. Kidera, "Supercritical boiler technology matures." Available at: http://www.hitachipower systems.us/supportingdocs/forbus/hpsa/technical_papers/CG2004.pdf

[20] T. Yano, K. Sakai, K. Kiyama, O. Okada and K. Ochi, "Updated low NOx combustion technologies for boilers, 2003." Available at: http://www.hitachipowersystems.us/supportingdocs/forbus/hpsa/technica l_papers/Mega2003.pdf

[21] H. Kimura, J. Matsuda and K. Sakai, "Latest experience of coal fired supercritical sliding pressure operation boiler and application for overseas utility." *Power-Gen Europe* 2003, May 6-8, (2003) Track 3: http://www.bhk.co.jp/english/ 4tech/contents/pge 2003 paper_blr.pdf

[22] K. Okazaki and T. Ando, "NOx reduction mechanism in coal combustion with recycled CO_2." *Energy,* 22 (1997) 207-215.

[23] B. Arias, C. Previda, F. Rubiera and J. Piss, "Effect of biomass blending on coal ignition and burnout during oxy-fuel combustion." *Fuel,* 87 (2008) 2753-2759.

[24] C. K. Man, J. R. Gibbins and K. L. Cashdollar, "Effect of coal type and oxy-fuel combustion parameters on pulverized fuel ignition." *2007 International Conference on Coal Science and Technology,* Aug. 28-31, Nottingham, UK, (2007).

[25] C. R. Shaddix and A. Molina, "Particle imaging of ignition and devolatilization of pulverized coal during oxy-fuel combustion." In: Thirty-Second Symp. (Int.) on Combustion, Pittsburgh: The Combustion Institute, (2009) 2091-2098.

[26] T. Suda, K. Masuko, J. Sato, A. Yamamoto and K. Okazaki, "Effect of carbon dioxide on flame Propagation of pulverized coal clouds in CO_2/O_2 combustion." *Fuel,* 86 (2007) 2008-2015.

[27] M. Taniguchi, H. Kobayashi and S. Azuhata, "Laser ignition and flame propagation of pulverized coal dust clouds." In: Twenty-sixth Symp. (Int.) on combustion, Pittsburgh: The Combustion Institute, (1996) 3189-3195.

[28] M. Taniguchi, H. Kobayashi, K. Kiyama and Y. Shimogori, "Comparison of flame propagation properties of petroleum coke and coals of different rank." *Fuel,* 88 (2009) 1478-1484.

[29] T. Kiga, K. Makino, K. Okushima, T. Abe, T. Hirano and T. Takagi, "Development of IHI wide range pulverized coal burner." Ishikawasjima Harima Giho, 27 (1987) 333-338, in Japanese.

[30] O. Fujita, K Ito, T. Tagashira and J. Sato, "Measurement of flame propagation speed coal dust using microgravity environment." HTD-vol. 269, Heat Transfer in Microgravity, *ASME*. 1993, (1993) 59-66.

[31] K. Kiyama, T. Tsumura, N. Sei, M. Taniguchi, S. Nomura, "Combustion technology for high fuel ratio coal." *Proceedings of International Conference on Power Engineering-97* 2, Tokyo, (1997) 477-481.

INDEX